# Preparing Artwork for Reproduction

David Cherry

Crown Publishers, Inc.,
New York

First Published 1976

Printed in Great Britain

**Library of Congress Cataloging in Publication Data**

Cherry, David.
Preparing artwork for reproduction.
Includes index.
1. Pictures—Printing. 2. Commercial art—Printing. 3. Printing, Practical—Layout.
I. Title.
Z257.C47 686.2'8 76-4802
ISBN 0-517-52618-2

# Contents

# Acknowledgment

I should like to thank Ted Baughan of Artco for photography and help; Dave Carter for retouching certain photographs; Focus Photosetting Limited for examples of work; Grant Projectors Limited for photograph and information; Halco-Sunbury and Co Limited for photograph and information of Copy Scanner, Letraset Limited for permission to display certain items; The Moby Dick Restaurant for permission to reproduce menu card (scraperboard artist: Biro); Brenda Naylor for some drawings; Ogilvy, Benson and Mather Limited for permission to reproduce John Player and Sons advertisement (scraperboard artist: Tony Gilliam); Royal Sovereign Limited (Magic Markers) for information; T A P and George Cumings for use of alphabet; Winsor and Newton for information; and my daughter for typing the original manuscript.

**David Cherry**

# Foreword

A vast amount of artwork is prepared for the printer every day. With the ever-increasing prices of printing it is all the more important to be able to mark up the material with clear instructions as to what is required and therefore to know something about printing processes.

For the commercial artist, whether in advertising or publishing or indeed anywhere where material needs to be printed, as well as for the art student, here is a guide to the technical 'know-how'.

I have compiled this book to share with others the knowledge which I have gained through experience, and to pass on to them hints and 'tricks' of the trade which I have found both helpful and time-saving. Obviously there are other ways of doing things as quickly and as well which the artist himself may discover and of which I should be pleased to learn.

It is, of course, impossible, in a book of this length, to deal with any one subject to any great extent. The business of commercial art is very large and complex. This complexity causes a tendency for people to specialize in one particular aspect.

The advertising part of the business in particular has undergone many changes in the last thirty years or so. For instance, many agencies now put out all of their finished artwork to outside studios and freelancers. This requires some sort of link to be set up with the agency and the freelancer, preferably a person with a good all round knowledge of commercial art besides having a good head for business. This person is called an 'Art Buyer'.

It is the art buyer's job not only to purchase artwork from outside sources and negotiate the price to be paid, but also to keep abreast of all the new techniques and to keep the 'creators' in the firm up to date on these matters. Such a person rates very highly in importance with all outsiders.

Going back to what I said earlier about specialist services, it is not uncommon to find not just individuals specializing but even whole studios who will work on only one particular branch of art. I am thinking in particular of studios that specialize in photographic retouching.

I personally specialize in lettering and calligraphy but do not do this work all the time. Occasionally I do design and paste up work. It certainly pays to be able to turn one's hand to most things and to have some knowledge of other aspects of commercial art.

To show the necessity for this knowledge, I have drawn up the following list. On one side I have put the people employed on the creative side of the business and on the other, the people who work from the agency supplied rough design and brief, to produce the finished art work.

| **Creative** | **Artist craftsmen** |
|---:|---|
| Visualiser | |
| Designer | |
| Layout man (or woman) | |
| Typographer | |
| Photographer | Photographer (copy negatives, prints) |
| Illustrator | Illustrator |
| Letter/Calligrapher | Letterer/Calligrapher |
| | Photographic retouchers – Black and white |
| | Photographic retouchers – Negatives |
| | Photographic retouchers – Colour |
| | Photographic retouchers – Transparency |
| | Photographic retouchers – Airbrush |
| | Paste up and key line |

As you will see, some of these people have a foot in both camps, this is not uncommon. The photographic retoucher too will probably be able to do other things as well.

**David Cherry**
Surrey 1976

# Printing Processes

Before preparing artwork for reproduction it is important to find out the process which is to be used and therefore to know the basic technical differences in order to save both time and money as what may be the correct procedure for one may not work for another.

## Offset lithography

This is a process of printing from a plane or flat surface where the areas to be printed are on the same plane as blank portions.

It is based on the principle that oil and water do not mix. The ink image is transferred to the paper by offsetting it from the plate to the blanket and from there to the paper.

The plates, nowadays, are made of zinc or aluminium sheeting, which can be curved around a press cylinder.

Complete artwork has to be assembled together as a 'paste up' with type matter – preferably with all portions in position ready for camera work. Artwork is referred to as *copy* by the lithographer and is described as either *line* or *continuous tone*. *Line* consists of solid areas or lines with no gradation of tone. (*Continuous tone*) or *halftone* would include such things as wash drawings or photographs.

'Repro' type (typesetting 'pulled' or 'proofed' on art paper) should be carefully mounted in position and must be absolutely clean and clear from smudges, as well as rubber gum, as these will pick up on the camera.

If a halftone is to appear with text on the same page, indication is made on the line portion of the artwork as to position and size of halftone.

The halftone should be scaled up or down, showing portion to be used either by marks in the margin of the picture or by a white paper mask cut to a scaled size.

After photographing separately, the halftone negative is 'stripped in' (printed down) with the line portion. When the plate has been completed and before going to press, a proof can be obtained on 'blueprint' paper as a means of final checking that everything is correct.

Where colours occur, separate proofs may be obtained on transparent paper overlays as a means of checking colour breaks, register of colour, etc.

## Letterpress

This is a process of printing whereby the inked plate or forme is pressed *directly* against the paper. The portions to be printed are raised metal pieces with the ink rollers touching only these raised parts.

There are three main types of presses: platen; flat-bed cylinder, and Rotary.

The blocks are generally made either from zinc or copper, depending on the subject, and on quality of paper to be used.

Artwork can be prepared similarly to litho or gravure artwork or supplied to the blockmaker as portions mounted up in the most economical way, i.e. at least 9 mm ($\frac{3}{8}$ in.) between items when reduced to reproduction size. This would apply only where portions are in proportion to each other. This minimum distance between items allows for the blockmaker's saw cut and sufficient on either side to act as a flange for pinning to a wooden base if so desired. Where this base is not required, instructions to the blockmaker would include 'unmounted'.

If mounted blocks are required, an accurate positioning guide on tracing paper must be supplied to the blockmaker for him to work from.

If the blocks are being supplied to the letterpress printer *unmounted*, then an accurate paste up can be supplied from blockmaker's art pulls or proofs, together with type mark up if this is required.

## Gravure

Gravure is an intaglio process, the image being etched or sunken below the surface.

Artwork for this method of printing may be presented in the same manner as for lithography.

Separate negatives are made of line and tone work which are then combined and printed out on clear film. The image is then transferred by means of a sensitized gelatine onto a copper cylinder. Where long runs of printing are required, the cylinder is very often chromium plated to increase its life expectancy.

When printing the cylinder revolves in a trough of ink and the etched portions are filled. Excess ink on the surface of the cylinder is removed by a 'doctor blade' (scalpel). When the cylinder contacts the paper, the remaining ink is transferred.

## Screen printing

Screen printing is used extensively on short or longish runs of such things as posters and showcards etc. It is a very versatile method, some of its important features being the ability to deposit a heavy film of ink therefore giving greater opacity than other forms of printing and on practically any surface.

Basically speaking this is a sophisticated stencil, that can be cut either by hand or prepared photographically, depending on the subject. The screen itself consists of a piece of open weave silk, metal mesh, nylon or similar material that has been stretched tightly over a wood or metal frame. The mesh holds the stencil firmly in place during printing. A squeegee is then used to force ink through the screen and stencil onto the printing surface. This can be done either by hand or mechanically.

Artwork may be prepared as for litho or gravure printing – but halftone work tends to be limited in quality because of coarseness of screens employed.

Generally speaking, good contrasting pictures must be used for reproduction as delicate tones cannot be satisfactorily achieved.

## Colour transparencies – reproduction

These should be supplied to the platemaker or blockmaker in as practical a size as possible to the reproduction size. It is asking a great deal from a 35 mm transparency to be 'blown-up' to a reproduction size of more than five times as several factors begin to show, for example: the slightest movement or poor focus become immediately apparent and the colours lose a great deal of their depth.

Wherever possible use 57 mm ($2\frac{1}{4}$ in.) square or 127 mm × 102 mm (5 in. × 4 in.) transparencies, marking the area to be used on the protective covering bag with a white or red chinagraph pencil.

Never finger transparencies, as marks will show up on reproduction.

Where the 'cropping' area is critical it is advisable to supply an accurate simplified tracing or photoprint to size of a major component. This gives the camera-man something tangible to work his proportion from.

When a set of letterpress four colour blocks have been made, they are proofed as separate standard colours i.e. yellow, magenta, cyan and black, usually in that order. A set of 'progressive pulls' are then made, beginning with yellow and overprinting with magenta. The next 'pull' contains the cyan and the final 'pull' contains all four colours.

These 'progressives' enable the blockmaker and client to check on densities and make certain corrections, such as reducing depth of colour, or highlighting specific parts thereby ending up with a more balanced look to the finished article.

The colours for each of the blocks or plates are separated by means of 'filters' placed on the camera lens, the only colour that cannot be 'filtered' out being the 'black'.

Nowadays, colour work is done on a high speed electronic colour scanner that contains a computer which automatically calculates the correct density, brightness etc. this being due to the computer calculating the size and density of the dots making up the halftone screens and where they overprint each other.

## Monochrome blocks

*The line plate*
A line plate is usually etched on zinc. First, a photo-negative is made from the original artwork. The negative is then placed face down on the metal plate which has been coated with a light-sensitive emulsion. Then both are placed in a photocopying frame and exposed to a powerful light. The metal plate is then developed and the image becomes acid resistant.

The plate is then placed in an acid etching bath and in varying degrees the unwanted parts or 'bites' are taken. When sufficient 'bites' have been taken, the plate is rinsed and dried and powdered with 'dragons blood'. After one powdering which gives protection, the plate is *burned in* to melt the resinous powder. When cool, it is powdered again – four times in all. The plate receives a final etch and is now ready for printing from. Care is of course taken not to undercut the relief portions as they would suffer badly and break down easily under printing pressure.

The other method, and in more common use today, is that of the *powderless etch*. This is a one stage process with the plate being placed face down on a turntable, which is then revolved, simultaneously etchant is splashed on the surface of the plate. The four way powdering is dispensed with as the etchant contains special agents that protect the walls of the printing image.

This method applies to both line or tone blocks.

*The halftone*
A halftone is made similarly to a line block but with the exception of a screen being used when making the photographic negative.

This 'screen' is placed between the lens and film, and can vary in coarseness from 22 to 60 lines per centimeter (55 to 150 lines per inch). These lines are laid horizontally and vertically. The square spaces between these lines break the image up into graded dots, varying in size and density according to the tone they are to reproduce.

Screen is determined by quality of paper to be used. Halftones in newspapers are generally 22 to 34 (55 to 85) screen. Smooth machine finished or calendered papers use 40 (100) screen and 48, 54 or 60 (120, 133 or 150) screens would be used on coated papers.

*Quarter tones*
This is a halftone that is reproduced by line method as follows: the tone photograph is supplied to a blockmaker who is requested to provide a screen print of the desired coarseness or fineness; generally it is coarse in texture. He places the tone photograph in front of the camera in the normal way and by means of his screens produces a screen negative. This is then placed in the enlarger and usually enlarged to twice or more times the reproduction size. This is to facilitate ease of working on by the artist. The print is then mounted in the normal way on mounting board and worked on with lamp black and process white. It is an exacting operation on the artist's eyes and only short periods of such work should be endured. The advantage of quarter tones is that where lettering is required such as on bottle labels, this can be cleanly picked out from the halftone background and made into what is eventually a line illustration. When the work has been completed by the artist, it is sent off to the blockmaker in the normal way with instructions for a fine line zinco to be made and reduced at the same time to the required size. Obviously at the outset, care must be taken to ensure that the correct coarseness of dot is ordered from the blockmaker which, of course, can be checked either with the magazine or newspaper, or from *British Rate and Data* (BRAD) in Great Britain and The National Guide to Media Selection in the USA.

The tools and materials of the trade are clearly shown in the diagrams and photographs on pages 14 to 17. Most of these are self-explanatory; the following need brief explanations.

# Equipment

## Pencils

It is a good idea to have a selection of grades of pencils or leads. The following grades cover most needs: 6B, 4B, 2B, HB and 2H. Some people prefer to use a pencil as hard as a 6 or even a 9H as a tracing down implement, but for this purpose an old steel knitting needle in a drop lead pencil holder is better as it does not need resharpening and gives a consistent line. Avoid using hard pencils on initial working out and planning, as this can leave a heavy imprint on the surface of the paper or board. Generally, people use hard pencils through sheer laziness, not wishing to be constantly sharpening the pencil. I prefer to use a drop lead pencil in conjunction with the sandpaper block, as opposed to the ordinary type of pencil.

Never allow the pencil to become blunt, particularly when marking out work, or it may lead to severe discrepancies in measurements. Sharpen the pencil on the sandpaper block by turning it over and over in a rotary motion back and forth, giving the tip a cone-like shape. This is generally a more useful shape than the chisel shape advocated for sketching.

## Pens

It is necessary to have a selection of pens, both for mechanical and freehand work. Some of the most useful, particularly for ruling up purposes, are those which have an ink cartridge in the body and a parallel tube with an inner needle of a specified thickness, which is measured in millimetres. These are particularly good in repetitive ruling up jobs, giving a constant thickness and quality of line. They must be kept scrupulously clean and not allowed to dry or cake, otherwise they quickly cease to operate and can be difficult to clean. It is possible to buy an aerosol into which the nib section can be screwed, and the cleaning fluid is forced through the tube, carrying away the offending particles of dried ink.

Square cut lettering pens of varying sizes will be found useful for the odd piece of calligraphic style lettering that one may be commissioned to do. These, of course, are used in the standard pen holder and there is on the market a more sophisticated pen with interchangeable nibs, that takes several types styles and sizes of nib. Be sure to specify for right-handed or left-handed person. A small mapping pen is useful for such things as stipple work or small lettering.

Pen lettering for presentation book

Miss E. Green, M.B.E.

in token of their gratitude and warm appreciation
of seventeen years of Loyal and Devoted service
in the interests of the Association and its affairs.

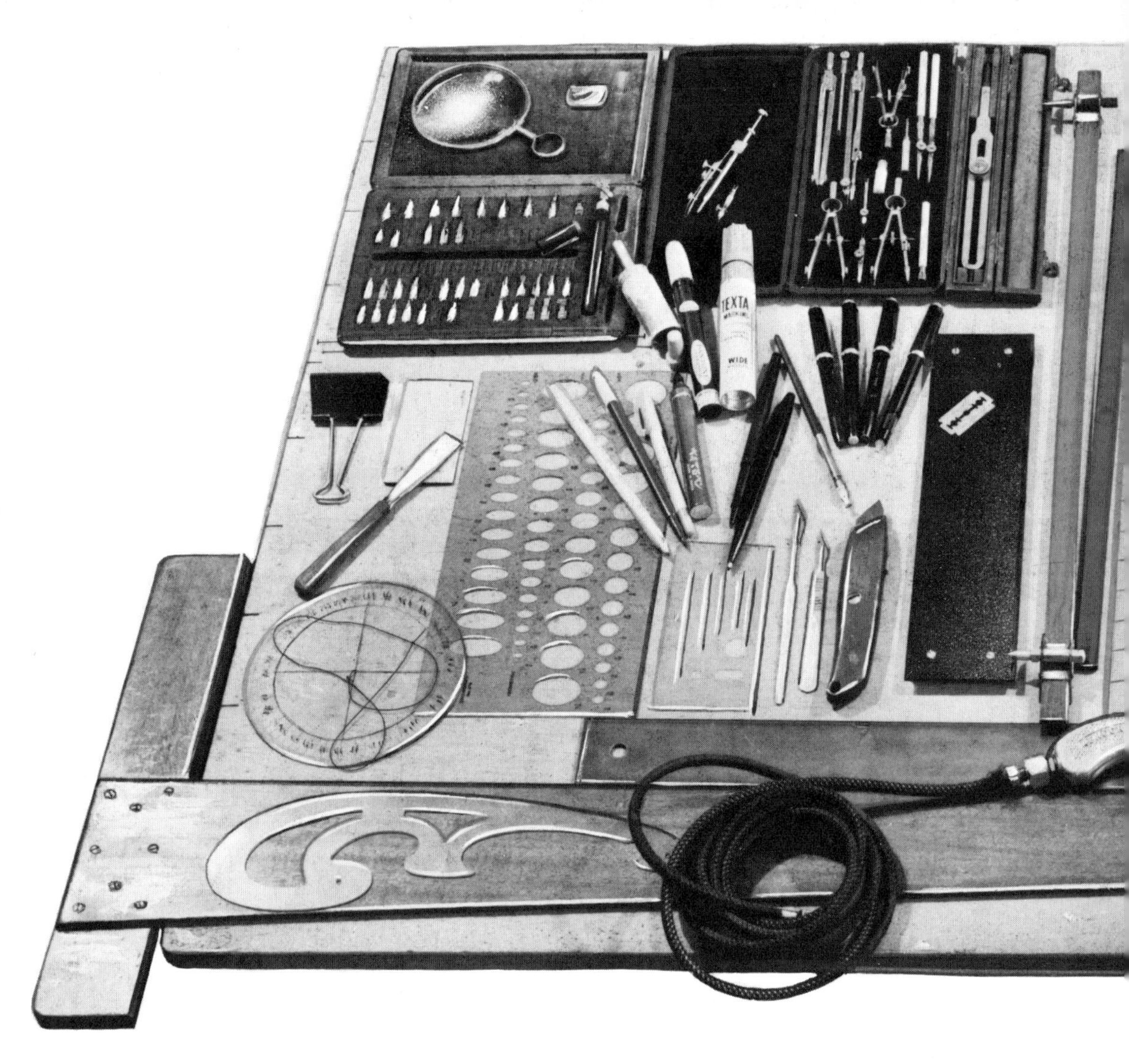
TEXTA
WIDE

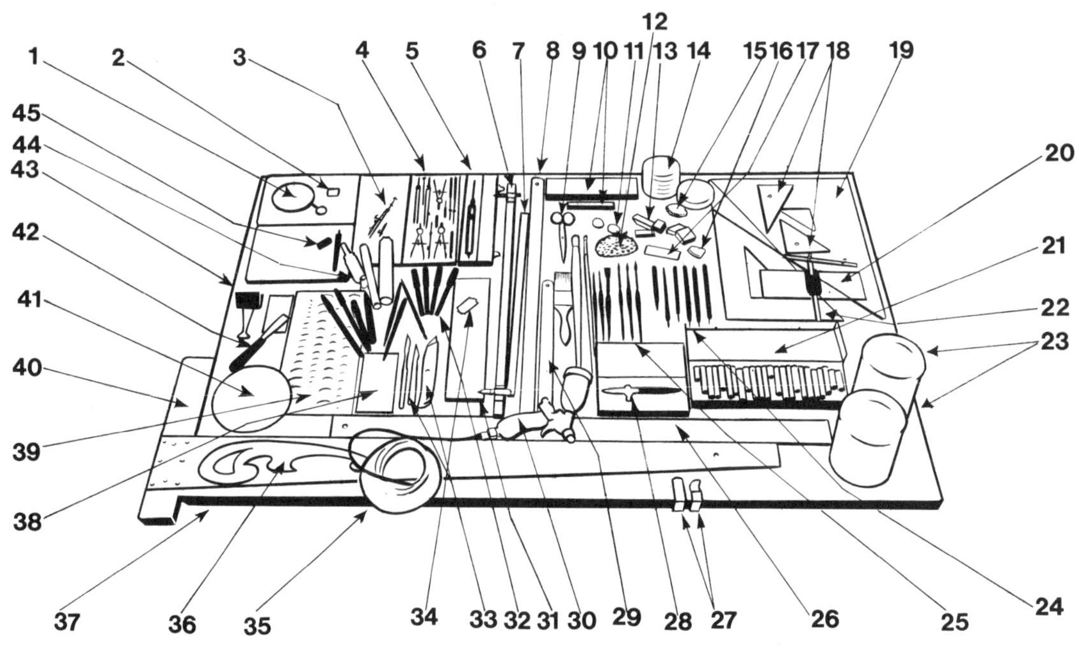
12
1 2 3 4 5 6 7 8 9 10 11 13 14 15 16 17 18 19
45
44
43
42
41
40
39
38
37 36 35 34 33 32 31 30 29 28 27 26 25 24
20
21
22
23

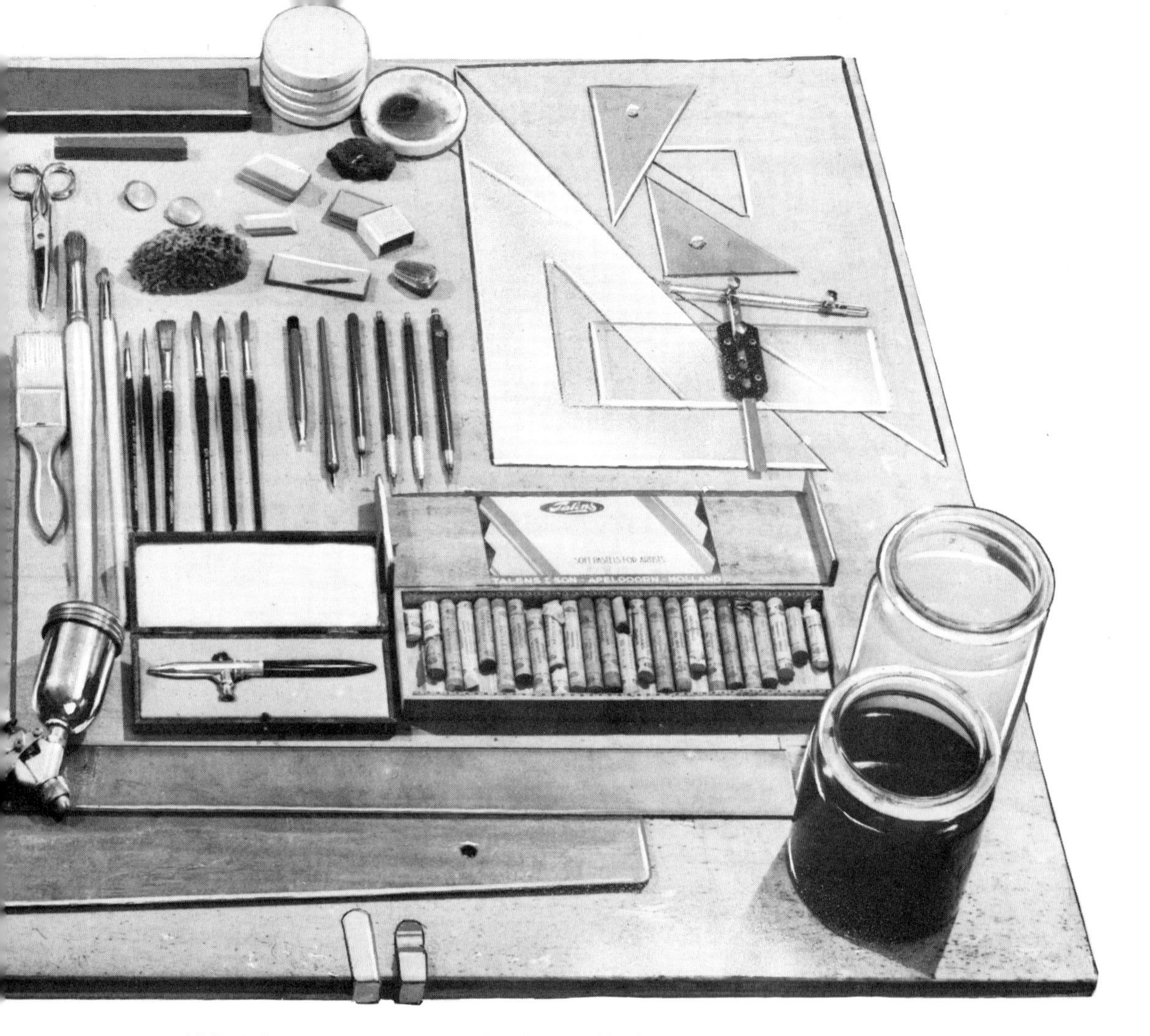


1 Magnifying glass
2 Reducing glass
3 Pump or drop compass for small circles
4 Set of instruments comprising: spring bows, pen and pencil compasses, dividers and two ruling pens
5 Proportional dividers
6 Beam compass
7 Wooden brush ruler or finger stick
8 2 ft (60 cm) stainless steel ruler
9 Scissors
10 Sharpening stones
11 Drawing pins or thumb tacks
12 Sponge
13 Assorted rubbers
14 Nest of palettes
15 Cow 'bungie'
16 Pencil sharpener
17 Sandpaper block
18 Small 45° and 60° set squares
19 Large bevelled 45° set square
20 *Rulalipse* – an instrument for drawing ellipses
21 Box of pastels
22 Large bevelled 60° set square
23 Two water pots
24 Fibreglass eraser
*Letraset* burnishing tool
Tracing down implement
Pencils in various grades
25 Brushes – Flat and round
26 3 ft (91 cm) steel straight edge
27 Clips for holding paper or board (instead of drawing drawing pins)
28 Airbrush or spraygun
29 Typographic point ruler
30 Large airbrush or spraygun
31 Hand bridge or support for working over wet areas without touching them
32 Rapidograph pens
33 Knives and scalpels
34 Razor blade
35 Air line for airbrush
36 French curves
37 Drawing board
38 Erasure shield
39 Small ellipses stencil
40 T Square
41 Protractor
42 Small square cut palette knife and piece of *Formica* for spreading rubber gum
43 Bulldog clip
44 Varying sizes of magic markers and felt tipped pens
45 *Graphos* pen holder and nib case

MIST
GILLAC
3-IN-
JOHNSON
BROWN
Triple Strength
PHOTO COLOUR
JOHNSON
BLACK-
SPOTTING
JOHNSON
LIQUID OPAQUE
PROCES
BLAC
103
LETRA
MATT
Sellotape
BARSOLA
PASTE
ENGLAND

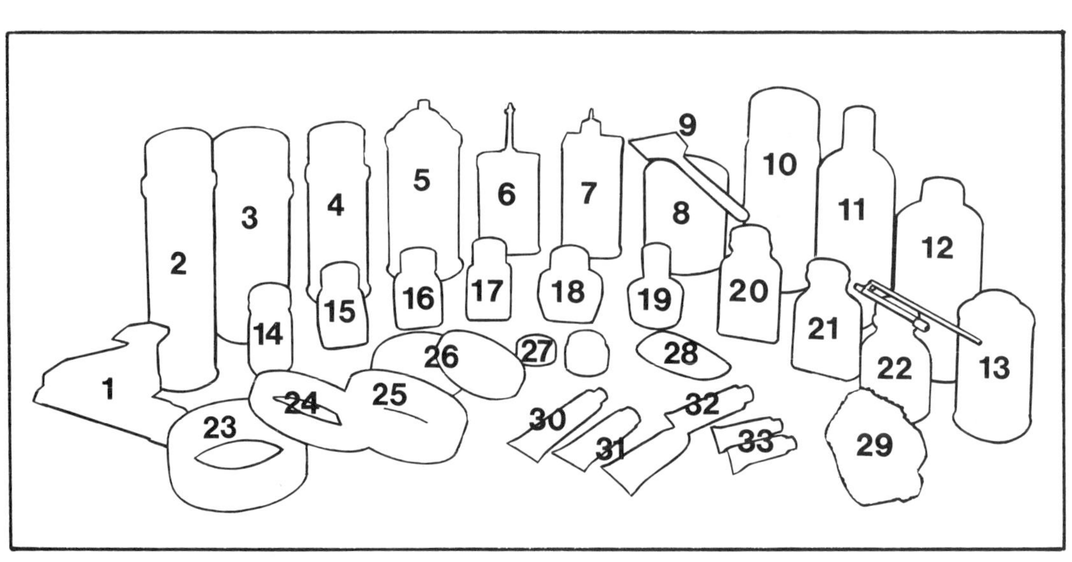
1
2
3
4
5
6
7
8
9
10
11
12
13
14
15
16
17
18
19
20
21
22
23
24
25
26
27
28
29
30
31
32
33

1 Stapler
2 *Letracote* spray 103
3 *Mistic* adhesive – a pressure sensitive adhesive for polyester or acetate film and other smooth surface materials
4 *Letraset* 101 protective spray
5 *Gillac* protective spray
6 3 in 1 oil
7 Lighter fuel
8 *Cow* gum (rubber cement)
9 *Cow* gum spreader
10 Fixative spray
11 Indian ink
12 *Gilstat* anti-static spray
13 Pounce powder
14 Process black
15 Liquid opaque
16 Photo-dye
17 Black photo-spotting dye
18 Black fountain pen ink
19 Coloured inks
20 Paper varnish
21 Gum arabic
22 Fixative and mouth diffuser
23 1 in. (25 mm) wide clear tape
24 $\frac{1}{2}$ in. (13 mm) wide clear tape
25 1 in. (25 mm) wide two-way tape
26 Barbola paste
27 Ox gall (solid or liquid)
28 Soap
29 Cotton wool
30 Special non-bleed white
31 Retouching greys
32 Gouache colours
33 Watercolours

### Erasers

These come into three categories: safety razor blade, glass fibre stick and hard rubber. It is best in the case of razor blades to have the two edged flexible type, rather than a single edged blade, reasons being that first of all, the two-edged blade has twice the life of the single edge blade and, because of its flexibility it is less likely to tear the surface of the paper. When using a razor blade to erase mistakes or marks from paper or board, it is best to apply pressure one way only, not scraping back and forth. Obviously, if a mark to be removed has gone a long way into the paper, it may be necessary to really work back and forth at this with the blade, finishing off with a one way movement to lay the fibres of the paper in the same direction. The same will apply to a fibre tip pencil. This must be used very gently indeed, as its action is extremely rapid and fierce. Often as not, it is best to do initial scraping out with an eraser pencil and finish off with a good sharp razor blade. Where neither of these happen to be at hand, one can sometimes use fine glass paper called flour paper. But where glass paper is used, care must be taken to ensure that it is spotlessly clean, or it will work dirt into the surface of the paper and leave a permanent stain.

### Rubbers

Three types of rubber are useful in the studio: the kneaded or putty rubber, the ordinary soft type of rubber and the pencil shaped type-writer rubber. Kneaded rubbers are particularly useful in highlighting pencil sketches, as they can be pushed into any desired shape. Their one big advantage over other sorts of rubbers is that they do not leave a residue on the paper or board. They are of particular use for removing multi-construction lines where a softish pencil has been used.

The second type, the soft rubber, is more use where a harder type of pencil has been used.

The typewriter rubber is of use where precise rubbing out is required or where lines are in hard pencil or indian ink.

A soft type of brush, like a baby's hair brush or a raffia brush is useful as a remover of rubber residue from the paper or board, rather than a cloth or handkerchief which only picks up and deposits the rubber elsewhere on the job.

Brush script

fibre look

### Brushes

Brushes most suitable for use in the commercial art studio are those labelled as finest sable hair. Size is largely a matter of choice, depending on the type of work to be undertaken. It is a mistake to buy brushes that have only short hair as they do not last long and, when worn, the quantity of ink or paint they can hold is very limited. Therefore, I would suggest that the type of brush to use is a long haired sable; size 4 or 5 for lettering or detailed work on wash or line-drawings.

When buying brushes one should always ask for a dipper or cup of water with which to test the brushes. Do this by taking about half a dozen brushes in the hand, dipping them in the water and then shaking them just once to get rid of most of the moisure. Brushes that are not suitable will immediately show by the hairs splitting. They should point and look firm. Having determined which brushes are suitable for further inspection, dip each one individually into the water and then work it about in circulatory motion on the palm of the hand to see if it springs back into correct shape. Always buy a minimum of two brushes at one time; one for white and one for black. Where indian ink is to be used, it is a wise precaution to have distilled water available in which to give the final rinse before storing the brush away.

Old brushes that have lost their fine tip can be used for brush scripts.

It is necessary to have a large soft type of brush (25–50 mm: 1–2 in.) for putting down washes or large areas of body colour. An ordinary house painting brush will serve very well or a wide oil painting bristle brush with a square cut end.

### Felt tip pens

These come in varying sizes and shapes. Some are of the re-loadable type with interchangeable nibs. The other types vary from the ordinary pocket pen size, with varying degrees of hardness of tip (useful for rough lettering and indication of type on layouts) to those with square-cut felt tips such as *Magic Markers*. These are of great use in covering reasonably large areas of flat colour or, when turned over, can be used successfully on visual aid type of lettering. If an extra large area is to be covered with flat colour, it will be quicker to take the marker apart and use the inner portion.

Where mistakes occur with felt pens, an application of rubber gum on the mistake immediately after the mistake has been made will help reduce the density of the colour when the rubber gum is removed. Better still, apply, with cotton wool in a circular movement, a small quantity of a solvent such as *Xylene*.

Felt tip lettering used on visual aid board

Tourist

Europe

## Airbrushes

Airbrushes are expensive but necessary pieces of equipment in the average studio. They can be used for spraying large flat or graduated backgrounds, retouching photographs of a mechanical nature or, with the nozzle removed, as a swift means of achieving a stippled effect. The actual airbrush can vary in size and pattern as can be seen in the illustration. With an expensive piece of equipment such as this, care must be taken to use water-soluble paints, except in the case of the large airbrush which can even be used for cellulose work. Extreme care must be taken in cleanliness after use. On no account should indian ink be used in the smaller type of airbrush as the needle, where it enters the nozzle, is extremely fine in diameter; in fact, if one compares it with a 5 amp electric fuse wire, it will be seen that the aperture is far less than this. The method of supplying the air to the airbrush can be achieved in three ways:

1 By means of a car type foot pump pumping air into a cylinder approximately 60 cm (2 ft) high and 20 cm (8 in.) in diameter. This is a rather long and physically tiring method, where a great deal of air is required. It is not recommended for large studios as it can be irritating to other members in the studio, particularly if the floor is springy.

2 By means of an electric motor with a piston in a small cylinder pumping air into the air cylinder. Today, these generally have an automatic cut out device when reaching a specified pressure, usually about 18 kg per sq cm (40 lb per sq in.). Cylinders on this type of machine can vary from 60 cm × 20 cm (2 ft × 8 in.) up to approximately 122 cm × 30 cm (4 ft high × 10–12 in.) in diameter. The cylinder is generally fitted with an automatic blow out device in case of failure by the electric cut out on the pump. With cylinders of this latter size, as many as six air lines can be tapped off quite successfully. Where multi-lines are to be employed it is best, if possible, to have the machinery in another room to save the constant and irritating pumping sound.

3 By means of a small portable aerosol container. These are probably best used only as an emergency measure as they can prove expensive.

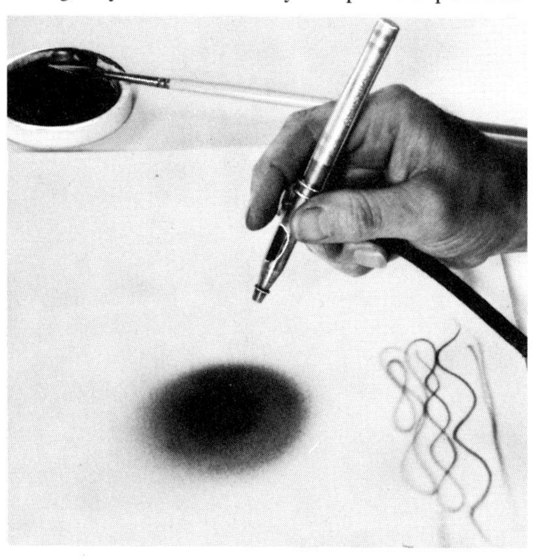

Airbrushing – note fine wavy lines that can be obtained on broad spraying effect

## Distilled water

A small bottle of distilled water can be obtained at any local chemist or garage (failing this, the water obtained from defrosting a fridge or freezer), and is the answer to any accident that may occur with indian ink, particularly where clothing has been splashed. Dab the affected area or work towards the centre, taking care not to spread it outwards.

## Cotton wool (absorbent cotton)

This should not be the medicated type both for reasons of absorbency and cost. It is useful for the photographic retoucher. A small ball wound round the end of a sharpened brush handle and wetted, is used for removing excess spraywork. Care must be taken not to get any in the bowl of the airbrush, as there is a tendency for one very fine hair to find its way down the needle orifice and jam the jet. It is also a useful means of rubbing excess pastel from sheets of tracing-down paper.

## Knives

A good heavy craft knife, such as a *Stanley* knife is useful where heavy board is to be cut. The best type of blade for use here is one with the single edge shown in the illustration which can be sharpened easily and rapidly. Care needs to be exercised in its use if the blade is of the retractable type. When using this type of blade, a good thick straight edge is advisable.

## Scalpels

These can be of particular use for retouching photographs or cutting up type matter for paste up work, and for cutting out intricate montages. The general shape of blade is the straight taper. Care must be taken in removing and replacing blades. It is advisable to have a small pair of pliers at hand for this. All knife and scalpel blades can be sharpened as long as great care is taken. Sharpen first on one side working in a circulatory fashion then turn the blade over and work backwards on the other side. The cutting edge should be honed towards the stone. Generally speaking, scalpels should not be used for heavy cutting duties, as they are liable to snap and possibly fly off and strike someone or something with considerable force.

When used as a retouching implement, it is generally best to scrape in one direction, so that the surface of the photoprint being worked on does not get unduly roughed up. The tip will be found particularly useful for picking off minute black specks or very small areas.

The other type of blade (curved) is useful for photographic scraping, but should not be used for cutting with generally, as there is a tendency for the blade to wander when used with a straight edge, with possible detriment to the job or fingers of the user.

Keep the fingers well away from the knife blade

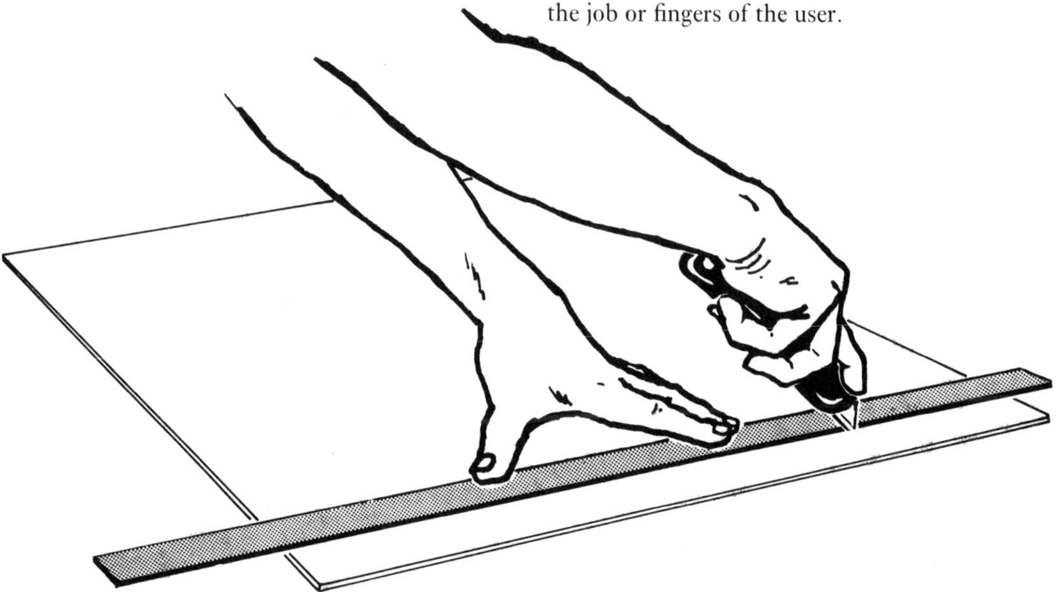

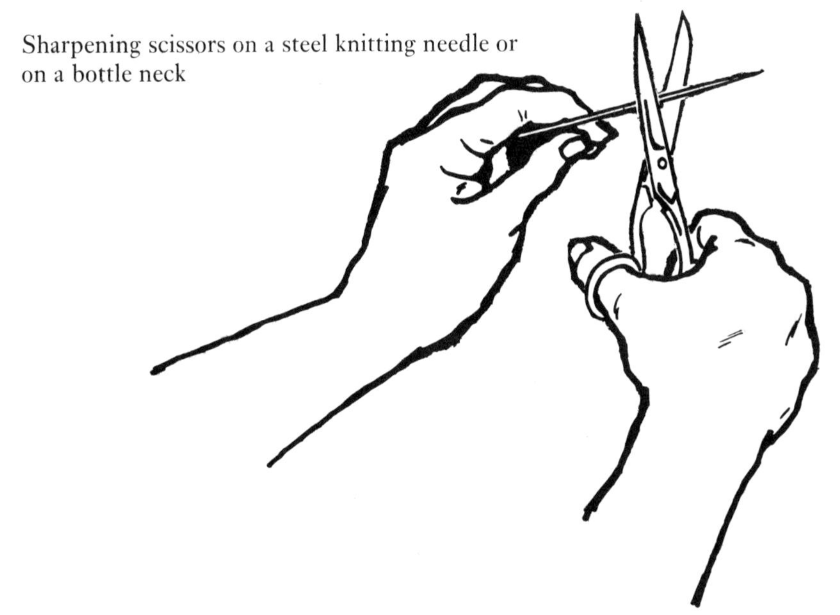

Sharpening scissors on a steel knitting needle or on a bottle neck

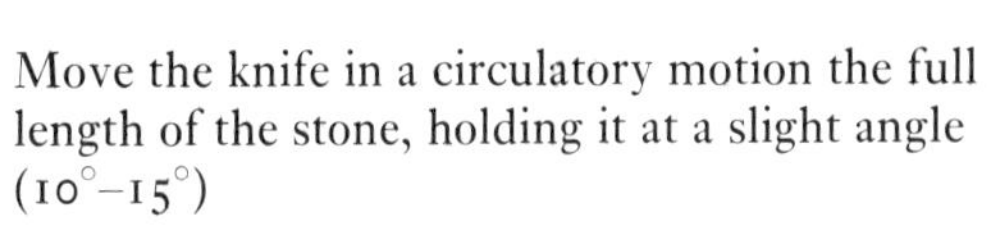

Move the knife in a circulatory motion the full length of the stone, holding it at a slight angle (10°–15°)

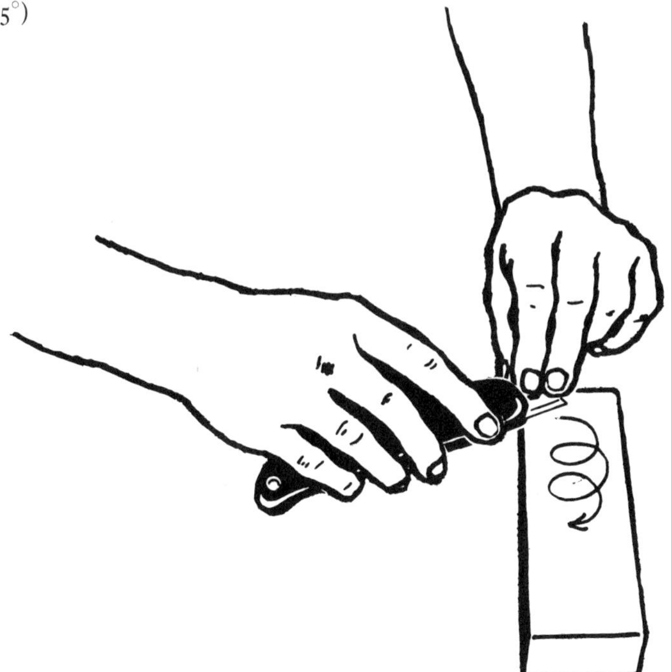

### Sharpening stone

It is a good thing to have a reasonable size carborundum stone in the studio with a small tin of fine oil. If this is not available a piece of slate may be used. This should be lubricated with water, not oil.

### Drawing board

A good drawing board with a true metal or ebony edge is essential. Generally speaking, the imperial size board 780 mm × 570 mm ($30\frac{3}{4}$ in. × $22\frac{1}{2}$ in.) is the most popular. If it is not intended to use a T-square on the drawing board, there is no need to go to the extra expense of buying one with the ebony edge. A good type of cheap drawing board can be satisfactorily made with a few simple tools and some 19 mm × 102 mm ($\frac{3}{4}$ × 4 in.) tongued and grooved timber, such as some flooring is made from, and then the top surface covered with a sheet of hardboard which can be pinned and glued. The only difficulty when using hardboard is that it does not readily take drawing pins (thumb tacks).

## Set squares

It is best to have a variety of angles and sizes of set squares for fairly obvious reasons; *viz* one 30 cm (12 in.) 45° set square, one 30 cm (12 in.) 60° set square. Personally I find that the 60° is generally more useful and less bulky to handle than its 45° brother as, when using in combination with a 60 cm (2 ft) ruler to cut or draw parallel lines, one can move the set square towards one's body for a much greater distance before the tip overruns the edge of the bench and gives one a sharp jab in the abdomen!

A thing that should be looked for whenever large set squares are being purchased is their squareness. This can be done as follows: borrow a straight edge in the shop and, with a sharp pencil, draw a vertical line with the set square and straight edge. Then turn the set square over and check it against itself, keeping the straight edge in its original position. Generally speaking these days, set squares are fairly accurate as they are cut by machine, but it is a wise measure to check anyway, not only for squareness, but also for flatness, the latter being a particular weakness in many 45° set squares.

It is more useful to have a bevelled set square than the flat variety, particularly where ink ruling is to be done. If, however, the reader has an unbevelled type, a useful tip is to affix a thin strip of card on the underside of the set square running about 15 mm ($\frac{1}{2}$ in.) parallel with the edge. This will be sufficient to create a small gap between the working surface and the underside of the set square and prevent ink runs underneath. It is also useful to have a miniature size of both the 45° and the 60° set squares, particularly for small work such as cutting pieces of type neatly and squarely.

## T-squares

It is useful to have two sizes of T-squares depending on the size of the board on which they are to be used. Personally, I use either a 60 or 90 cm (2 or 3 ft) ebony edge T-square, as these are durable, although more expensive than the plastic type which are rather 'whippy' and tend to bow when used. The ebony edged type can also be easily trued should it inadvertently become damaged along the edge.

## Rubber gum – rubber cement

Although this has been in use for many years in studios, the term 'paste up' is still employed as a hangover from the days when this was literally so. When using rubber gum it is best to spread a thin film over each of the surfaces to come together. A piece of stiff plastic, for example *Formica*, approximately 5 cm × 10 cm (2 in. × 4 in.) can be used as a spreader for medium size areas such as a 30 cm × 25 cm (12 in. × 10 in.) photoprint. For small pieces of type an oil painting palette knife with the ends squared off will be found extremely useful and not wasteful. It is a mistake to use large quantities of rubber gum: the waiting time before bringing the two surfaces together is lengthened; an excess of rubber gum does not stick so well, and it is wasteful. To save unnecessary waste, areas can be masked off with old off cuts of paper after marking out lightly with pencil. If, however, one wishes to move type around, one can afford to be a little more generous, and bring the two surfaces together whilst still wet. Further, should it be necessary to move the piece of type and when the rubber gum has dried off, a tin of petrol lighter fuel will be useful as a quick means of solvent. Care should be taken however with this as it is inflammable, and also can cause the ink of the type to smudge.

To remove excess rubber gum, the fingers can be used, but by far the most successful method is to use what is known as a 'goobungie', a 'Cow bungie' or a 'Cow rubber'. The latter names are derived from a particular manufacturer of rubber gum. This can be gathered in a very short time from around the edge of the tin, or by pouring a little out on a piece of paper and leaving it to dry off. When used, of course, the 'bungie' grows. It will be found necessary to peel pieces off at times, particularly if rubber gum has been used on painted surfaces, as some paint may pick up in the bungie and be spread on to the surface of the next job, with disastrous results. When removing rubber gum from acetate or *Kodatrace* surfaces, it is usually best to dab or roll the bungie, rather than a backward and forward motion, as this leaves friction marks on the surface.

There are several makes of aerosol rubber gum on the market that are useful, particularly where intricate cut-out jobs are to be affixed permanently to a background.

### Fixatives

These come in various guises; some of them may be used with a mouth diffuser and are suitable for pencil, charcoal and pastel work. The same type of fixative can also be purchased in an aerosol, which of course is a much more expensive way of buying. There are on the market several aerosol packs, such as *Gillac*, that will do a multitude of jobs. This can be used with perfect safety on most jobs, provided the instructions are carefully followed. The cans should always be shaken before use to make sure that the propellant and the fixative are well blended. Safety precautions must be strictly observed, according to the manufacturer. Not only can these aerosol fixatives be used on the aforementioned items, but they can also be used to create effect on sprayed backgrounds or photographs that have been retouched by airbrush, to give both protection and a gloss.

### Gum arabic

Gum arabic mixed with water of approximate proportions of 1 : 10 can be satisfactorily used through an airbrush as a means of protecting retouched photoprints and providing a gloss. The airbrush should be scrupulously clean and run through with warm water after use. It is also useful as a medium for helping to bind paint together on non-absorbent type surfaces.

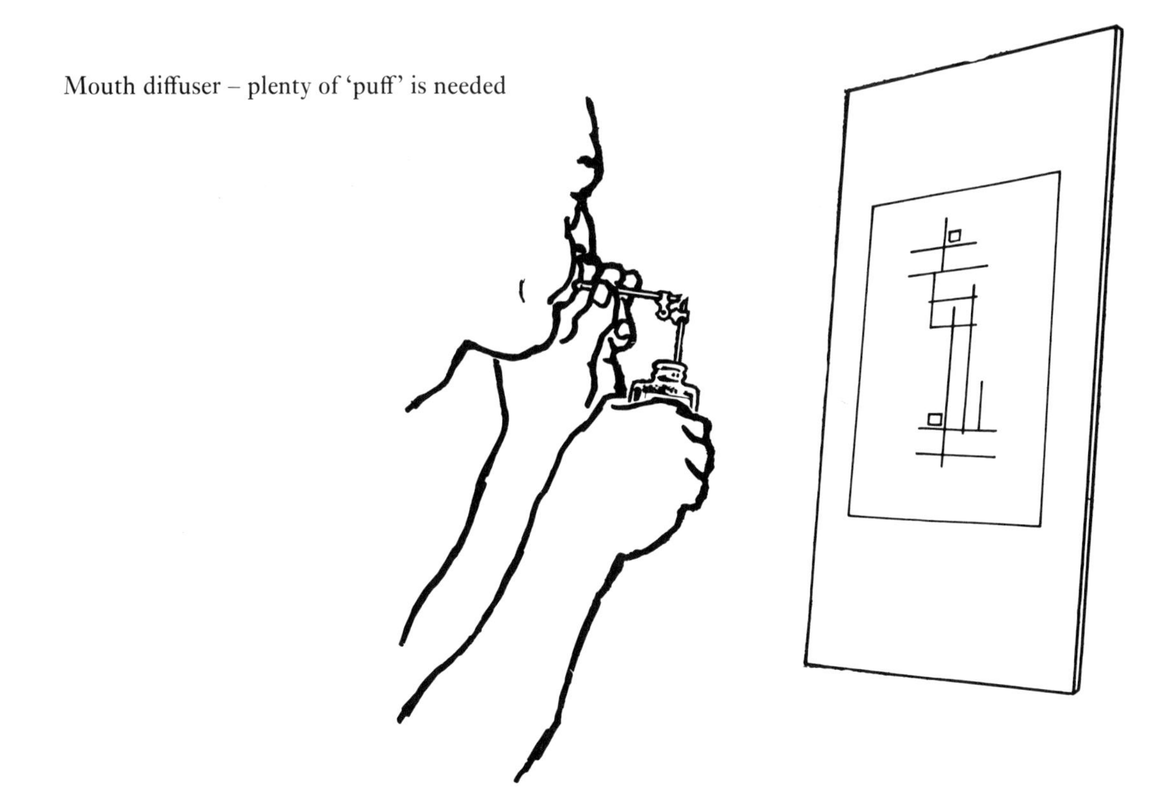

Mouth diffuser – plenty of 'puff' is needed

## Sellotape or Scotch tape

Useful not only as a means of taping tracing papers down, but also as a masking medium when painting, pastelling, airbrushing or using *Magic Markers*. The edge should be well burnished down when used as a masking medium otherwise it will, when used on rough papers, allow moisture to creep underneath it giving a rough edged effect. Where soft paper is to be used, to prevent the surface being roughed up when the tape is being peeled off, it is best to lightly rub the adhesive surface of the tape against the edge of a table. This removes a great deal of tackiness from the tape and still provides sufficient for it to adhere and do the job successfully, even on cheap type mounting boards or pastel type papers. It is also useful for stopping lines on multiple ruling jobs.

It is probably one of the most wasted materials in general use in studios. Novices and those who are not concerned with cost will probably use a piece of tape that is two or three times too long.

## Two-way (twin-sided) tape

This is an absolutely invaluable aid in the studio coming in for all sorts of jobs, one of the chief being mounting individual letters or numbers of repro type. It adheres much more readily and firmly than rubber gum, which has a tendency to pull out with the excess rubber gum when cleaned off from the surrounding area.

Where possible, two-way tape should be used on IBM setting in preference to rubber gum, as this has a tendency to remove the image from the surface should the two come into contact. The tape is also widely used for lightly tabbing down proofs of photographs or even cloth-like materials in such things as salesmen's kits and folders.

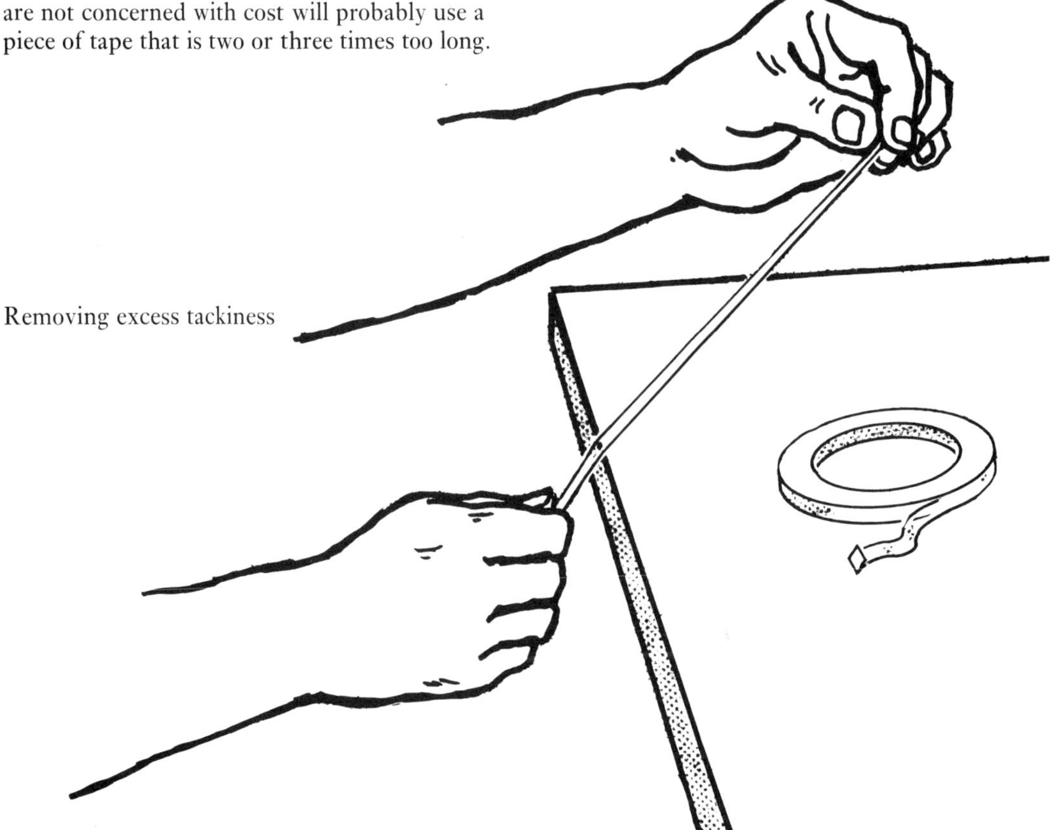

Removing excess tackiness

## Barbola paste

Barbola can be bought in varying sizes of tins. The average size used in the studio is roughly the size of a tin of shoe polish. It is a putty like substance to look at and is applied by wetting the brush and rubbing it over the surface of the barbola paste as one does with palettes of ordinary colour. It is applied to the surface of the paper like paint, its purpose being to raise a surface on the paper and give an embossed appearance to such things as letterheads, packaging that requires a raised surface or even manuscript work. By doing a series of coatings of the paste, allowing drying time between each one, a substantial embossed effect can be obtained, finally finishing off with a metallic colour, a plain colour or even white, depending on the desired effect. Where metallic colours have been used, the careful use of an agate burnisher can heighten highlights a great deal and give design an added dimension. By the use of ordinary white paint over the paste, a very good simulation of blind embossing may be obtained. Where ordinary water colours are painted over the paste, a little gum mixed in with the paint before application results in a very satisfactory resemblance to dye-embossed letterheads.

Gesso powder can be used in a similar manner, but can be mixed with water or paint.

## Pastels

A box of pastels in the studio can be of great help in preparing either roughs or occasional finished artwork. They can be useful for the loose type of sketches on roughs, where full colour is intended, and are particularly useful for covering large areas with flat colour over which lettering has to be applied: the one advantage here being that the paper does not cockle as it does with the application of paint. A pastel fixative will be needed, of course, which can be applied either by means of an aerosol or the old and tried method of a diffuser. Pastels are also extremely useful for use as a tracing down medium i.e. application to the back of thin paper and using as a pressing through agent.

## Paints

For most illustrative or lettering works, designers' gouache colours such as jet black, lamp black and permanent white will cover most needs. Jet black has a good coverage for large areas, but tends not to give a good sharp edge as is sometimes required in lettering jobs. It seems to lack the fineness and 'fattiness' that one can obtain from a lamp black. It is extremely good for filling in letters and giving a full velvety black look, highly suitable for reproduction. Permanent white is used for 'tickling up' letters or illustrations, as it does not 'pick up' under platemakers' lights. Permanent white has a better coverage than ordinary process white.

Gouache colours, of which there is a wide range, are generally used for full colour illustrations. They have good coverage and good density of colour suitable for reproduction.

Water colour paints are often used for tinting or air brush work, as they are very finely ground. Here I would add a word of warning: some colours, such as emerald green, may be highly toxic, containing minute quantities of arsenic. Therefore the rather dangerous habit of licking brushes is to be deprecated.

It is useful to have a tube of sepia water colour which can be used for retouching purposes on photographs. It gives a warmer and more accurate tonal match on photographic prints than ordinary black.

## Papers and boards

Depending on the type of job to be carried out, i.e. fine line work or wash work, then the choice of surface is of paramount importance. Generally, three types of paper or board come into use. The *hot press* type of board with a smooth sheen will be found to be of particular use to the lettering or general artist involved in key-line work, where fine sharp edges are required. Flat colour can be laid with a brush on this type of board, as long as the area to be covered is masked off accurately and is damped with clean water before applying the paint. A good type of hot press board should be able to take a reasonable amount of scratching out with razor blade where erasure is necessary.

Some inks and paints tend to slide across the surface and not lay very opaquely. This will be found out only by trial and error.

The *not surface* types of board are those generally used by illustrators, the matt surface being of use to the artist who is going to work in line and wash medium, the surface being readily acceptable to both pen and brush work. *Water colour* paper or board is generally very rough surfaced, but is useful as a dry brush medium. It is very expensive.

*Bristol boards* are a particularly hard double-sided thin card, used by pen artists, providing them with a particularly good surface for their work. It takes severe punishment from either razor blade or ink rubber. It is an expensive board and not often found in general studio work.

*Carton boards* are cheap, thin cards with a white surface on one side, suitable for model-making and, as the name implies, carton making. It is readily creased and folded and for the purposes of 'dummies'. Provided it is carefully handled, illustrations and lettering may be applied successfully.

## Synthetic overlay materials

The following is a short list of some of the materials in this category that are used for accurate second or more colour workings, or type for overprinting: *Kodatrace*, *Ethulon*, acetate, *Permatrace*, *Herculene*, *Astrafoil*. Care should be taken regarding correct use of these materials. For instance, there may be a warning 'Do not under any circumstances use rubber gum'. The reason is that the petroleum or the benzine spirit in the rubber gum tends to distort the material and permanently buckle it. In most cases indian ink or gouache colour can be applied successfully, but even so, before applying, it is wise to give the surface a wipe over with either clean water on cotton wool, or possibly methylated spirit, or lighter fuel. This dissolves any grease that may have adhered to the surface from the user's fingers. Another medium is 'pounce' which is like a coarse french chalk. This may also be used to kill any grease that may have adhered to the surface.

Where there is no 'tooth' on the material, such as *Astrafoil* or acetate, it may be necessary to use 'pounce' or mix the paint or ink with either soap or ox gall, but even so this has a limited life and tends to crack and break up if handled too frequently; also one has the problem of discoloration, particularly of white or light colours if ox gall is used.

Care must be taken when using these materials regarding accurate register, as contrary to most manufacturers' claims, it is only some of the polyester-based materials, such as UNO Drafting film, which can be guaranteed 100% foolproof regarding movement, ie stretch or shrinkage. Where two or more workings for colour are required, it is advisable to check with the plate or blockmaker whether he prefers to work from overlays of this nature or from keyline drawings.

## Projectors

There are several projectors of the camera type on the market. They vary tremendously in price and, generally speaking, are found in larger studios. Those fortunate enough to have one of these pieces of equipment certainly have a big time-saver. Generally, they are a box-like arrangement with a platform that is adjustable vertically, and a bellows set vertically above the platform, which is also adjustable. Degrees of magnification and reduction are dependent on the lenses that are supplied. One make of projector has a lens that will enlarge or reduce approximately five times; this of course is very useful but can be extremely inaccurate on the magnification end of the scale, due to distortion at the edges of the image. These projectors have banks of lights that are fitted in a reflector surrounding the lens and underneath the bellows, and may be cold cathode, fluorescent or ordinary light bulbs. The image is thrown on to a thick glass panel on the top of the box which may be clarified by means of tracing paper or some such semi-opaque medium. It is then possible to trace off fairly accurately the image required. These projectors are particularly useful where adaptation of artwork is required and the user wishes to find out how big component parts may be needed e.g. photographs and logo-types, areas of copy, illustrations, etc. Besides being able to enlarge or reduce artwork on projectors it is also possible to place a small light box on the platform and by means of the light from within the box, enlarge or reduce transparencies for tracing purposes.

In some of the older types of studios these projectors are commonly known as *Lucy*. This is derived from the old type of projector now very much out of favour and called the *Camera Lucidagraph*. This is a telescopic tube with a G-clamp (or C-cramp) at its base surmounted by a mirrored prism which is used in conjunction with a set of lenses varying in magnification and reduction, the object to be copied may either be an object in three dimensions or a flat object pinned to an upright copy board. The height of the prism and the distance of the prism from the object plus the lenses govern the size of the object's reflection on the working surface which is at right angles to the object itself. By turning the prism 90° the image may also be reversed from left to right. It is a very old form of projector and certainly proved invaluable in the past. Its big disadvantage is that the slightest movement of the telescopic arm affects the reflection and many people find it a physical impossibility to use such a piece of equipment.

Intricate cutout halftone of Grant Cold Cathode Unit

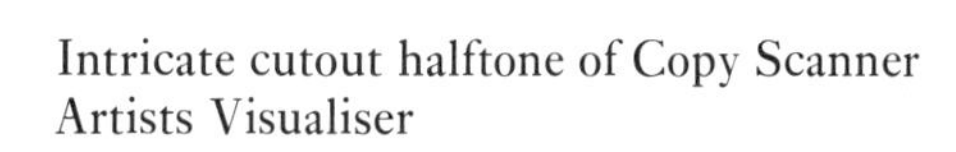

Simple cutout halftone of Grant Projector reproduced from photoprint

Intricate cutout halftone of Copy Scanner Artists Visualiser

## Flimsy

Flimsy, or to give it its proper name, 'imitation greaseproof paper', is best bought by the ream, as it is the most used of all the papers in the studio, not only as a means of tracing or sketching out things in a rough state, but also as a cover paper to protect artwork. It is ideal because one is able to write on it, referring to particular items on the artwork or copy, or instructions to the process man. It protects the artwork from odd splashes of water and dirt.

## Blotting paper

No self-respecting studio will be found wanting in this very basic of all papers, if for no other reason than mopping up the occasional bottle of ink or colour that may be inadvertently spilt on the job. It is even useful at times as a means of illustration medium, applying a dry brush technique to obtain effects that one may desire.

## Gum strip paper

Brown paper with one sticky side used for stretching paper to take colour washes. See page 41.

## Cover papers

These come in several thicknesses, colours and qualities. Often used to cover art work for presentation purposes, a colour often becomes associated with a particular studio. Their real purpose, of course, is not just top dressing, but as an added protection against knocks and rain splashes. They can be used for presentation roughs, acting as a background or frame. Black is the most commonly used, but neutral colours seem generally to enhance the work more. These papers are sometimes used for lettering or illustrating, the only fault here being that the surface tends to be somewhat hairy. Water must be used sparingly on it as it tends to cockle badly unless pre-stretched. Body colour, particularly white, takes well on these papers.

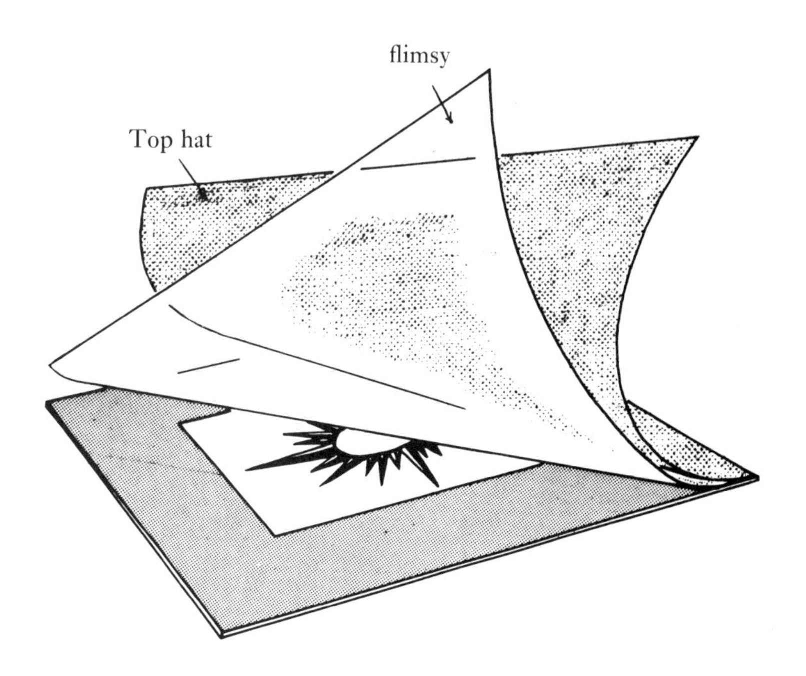

# Safety in the Studio

Considering the types of sharp implements in studios, there are remarkably few serious accidents. Nevertheless, they do occur occasionally and usually the fingers of the left hand (if one is right handed) are what suffer most. For example, when cutting through paper or board with a sharp knife in conjunction with a ruler or straight edge, the knife, instead of travelling along the edge of the ruler slips over the top and slices into the index finger. This is probably the most common of all accidents. Here we have to get our priorities in order. Obviously if the cut is very severe and bleeding is profuse, the job comes last and one's fingers first. Therefore, the golden rule must be a first aid box in the studio. If this is not the case and the bleeding is rapid, then a lump of ordinary cotton wool that one should have to hand for retouching purposes and general work, can be swiftly wrapped around the finger and tied up with *Sellotape*. One of the commonest causes of cuts in the studio is the edge of freshly cut paper. This can be exceedingly sore and very often deep. Care should be taken when handling paper never to run the tips of the fingers along the edge. In such instances, it is useful to have a styptic type pencil handy. This quickly stops any bleeding. As a general rule, when cutting with a ruler and knife, the edge of the ruler furthest from you should be used, with the knife pressed up firmly against it (see illustration on page 20). There is a definite lack of control if the nearer edge is used, the tendency being for the tool to come away from the straight edge and curve in towards one's body. Blood on the job can be exceedingly difficult to remove (if not impossible), but should it occur then dabbing with clean, cold water and cotton wool can help to remove most of the mess.

# Presentation Roughs

As the name implies, these are for presentation to the client, and are usually of a very high standard indeed, particularly when done as speculative work. Much care, time, expense, thought and planning can be put into work of this nature. Obviously the prime interest of the designer is to impress the would-be client with good quality work with logical, clever thinking. The standard of work generally needs to be high because of lack of knowledge and understanding on the part of the prospective client, as well as inability to understand rough visuals. In many cases the client expects to see work that is comparable with the finished article. Some designers even go to the lengths of not only writing the copy for the design, but even having the text set up in type and proofed on the actual rough artwork. Of course, exercises of this nature and depth can be extremely expensive and time consuming. All very well, of course, if the designer manages to land a new client with his impressive display. Generally speaking, this type of presentation is best left to the large studios and advertising agencies that can afford such luxuries. It is not unusual practice when presenting roughs for advertisements to obtain copies of the magazines and newspapers and actually mount the roughs in position on the page, thereby giving a much more effective appearance to the advertisement. The other method is to carefully mount the design on a suitable paper or board background with a margin all the way round which acts as a framework to separate it from the background. The whole is then generally mounted in a coloured cover. When the client becomes aware of the designer's method of working and thinking, he will probably quite readily accept a rougher form of design and be more interested in the copy and the idea behind the copy, particularly if it is technical in nature.

A good idea is to use photographs if possible for certain rough ideas and it can be extremely useful if one has access to a *Polaroid* type camera to use as a means of reference. In the case of rough visuals or 'scamps' as they are sometimes called, either very soft pencil, pen and wash or felt tip pen techniques may be used successfully, depending on the subject or preference of medium. Where pencil or felt tip pens are used, it is generally best and most easily worked on thin layout paper that one may procure in book form varying from A1 to A5 size. A word of warning here: always, when using a layout pad, insert a stout sheet of paper or card underneath the sheet being worked on, otherwise severe indentation, or bleeding through of colour, may take place for several pages through the book and be highly noticeable when worked on. It is usually best where pencil or perhaps pastel is used to apply a suitable fixative as a protection.

# Methods of presenting Artwork

When presenting artwork for approval by the client, it is general practice to put a flimsy overlay sheet on the artwork. This can be done by several means, the most common being with a flap over on to the back of the board and fixing with rubber gum or *Sellotape*. The function of the flimsy is for protection of the artwork and for the client to make notes regarding possible alterations in the appropriate places. Where photographs are mounted on the artwork, one must be careful to steer the client away from writing anything directly over the photograph as, generally speaking, the pressure from the pen or pencil is sufficient to go through the flimsy and dent the surface of the photograph. This is particularly noticeable where the photograph has been glazed and short of lifting the photo from the board and soaking it in water and re-drying, the dent cannot be removed with any great certainty and will most certainly be picked up as a reflection when put in front of the block-maker's lights and camera. Where corrections to illustrations are needed on photographic work, notes should be made on the flimsy at the side of the paper.

Hand drawn 'register marks' used on overlays to guide platemaker in positioning 'strip in' or 'second colour' working. Printed versions are obtainable

From the artwork the platemaker has made 4 separate colour plates then duplicated them 6 times on each plate. Note the bleed and register marks left on to help the printer in correctly registering his colours

Overlay 1 yellow

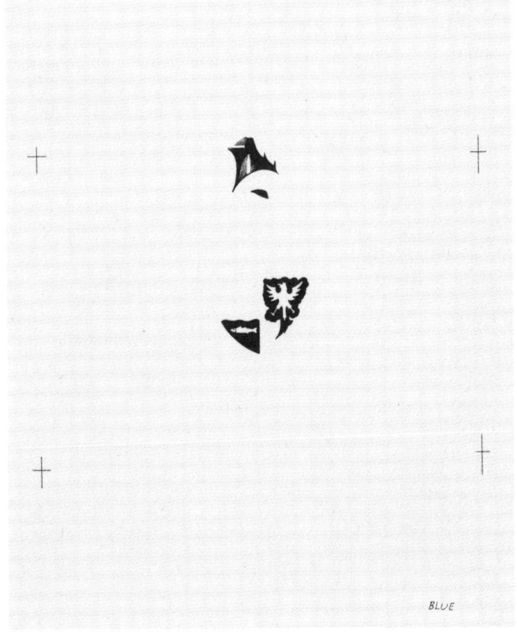


Overlay 2 blue

Overlay 3 red

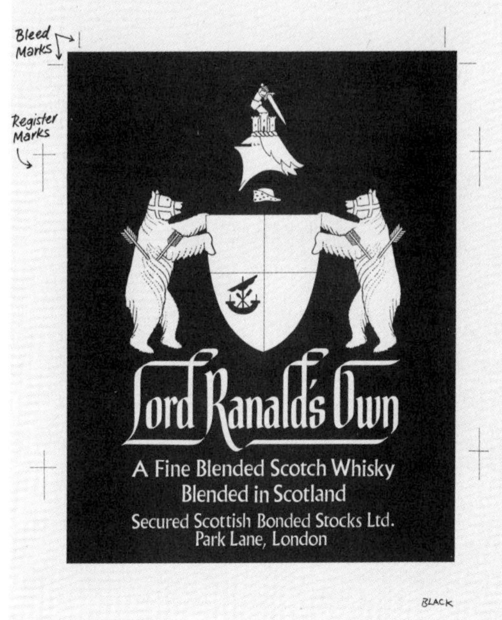


Flat artwork on board – black printing. Note the register marks and bleed marks

## Cutter drawings and bleeding off

Where a finished printed job is to be trimmed it may require, due to its shape, a special cutter being made. These are handmade and are very expensive items. They are used a great deal in label work such as wine labels, cosmetics, etc. It is usual practice to make a drawing of finished size so that the cutter maker can readily and accurately bend his metal. Generally speaking, cutter drawings are done on a *Kodatrace* type of overlay and drawn with pen and ink with very thin lines to indicate the area. Where a printed job will be 'bled off' it is customary to put four sets of corner ticks indicating the area to be trimmed. This only occurs where the colour of the printed image is to run off the page or area and as with cutter drawings, the accepted trimming off margin allowed is 3 mm ($\frac{1}{8}$ in.). This allows for any discrepancy that may occur during printing or trimming.

## Vignettes

These can be tackled in several different ways, depending on personal choice. Generally speaking though, they are done by means of an airbrush. In all probability they are the trickiest of all things with an airbrush. One of the biggest problems is keeping the colour flowing smoothly through the airbrush without spitting, particularly when the colour is very faint. The surface of the paper or board can affect the gradual fade away to a large extent, some papers tending to show up blotchy patches where the colour has been absorbed unevenly. The most difficult vignette of course is one where no edge must show. They should always be worked with the darkest portion nearest to the nozzle of the airbrush itself (see illustration on page 18).
the colour furthest from one is then obtained by a means of colour bouncing from the surface of the board and not directly attributable to the airbrush itself.
Vignettes should always be well protected with either gum arabic or some fixative or protective film, as they are very prone to marking. If fluff or any foreign body is dropped on a vignette, no one should ever blow with the mouth, as invariably marks will ensue from spittle. Neither should they be fingered. To remove foreign bodies it is best to turn the board on edge and just give it a slight tap on the edge of the table. It should be sufficient to remove any object. If it still persists in adhering to the surface, a pointed instrument such as a scalpel can be used with no ill effects, or even use of the airline from the airbrush.

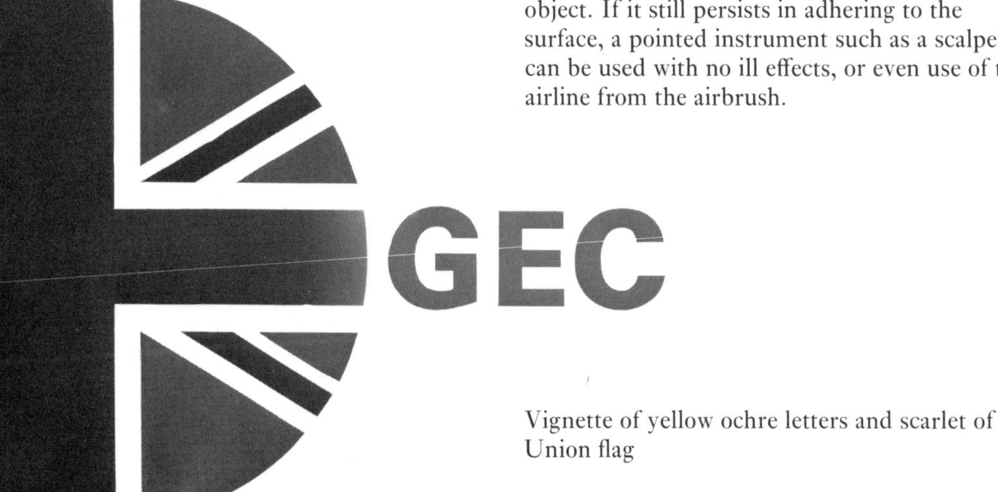


Vignette of yellow ochre letters and scarlet of Union flag

## Drawing cartouche or border

Sometimes a fancy border or a cartouche is required, and to save a great deal of time and money, the best method to work is by carefully drawing a portion of the border up in size and then ordering several photoprints of this to the size required, and stick them up in position on a piece of mounting board. As regards a cartouche that is perhaps in the form of a circle, oval or intricate shape, it is, generally speaking, best to draw a quarter or half of the finished artwork up in size, then order 1 or 2 prints to the required size and 1 or 2 prints again to the required size but reversed left to right, carefully leaving a centre point and some guide lines on the extreme edge as a means of accurate positioning.

Two part cartouche with one portion reversed left to right

## Pie percentage graphs

To divide up a circle accurately into % portions, two methods may be employed. One is the mathematical formula $\frac{x}{100} \times 360$ which gives the answer in terms of degrees that can be marked off accurately with a protractor. The other method is also by a simple mathematical formula and is worked as follows: 3.60 = 1%. This multiplied by any required percentage will give the angle required to be marked off, again with use of a protractor.

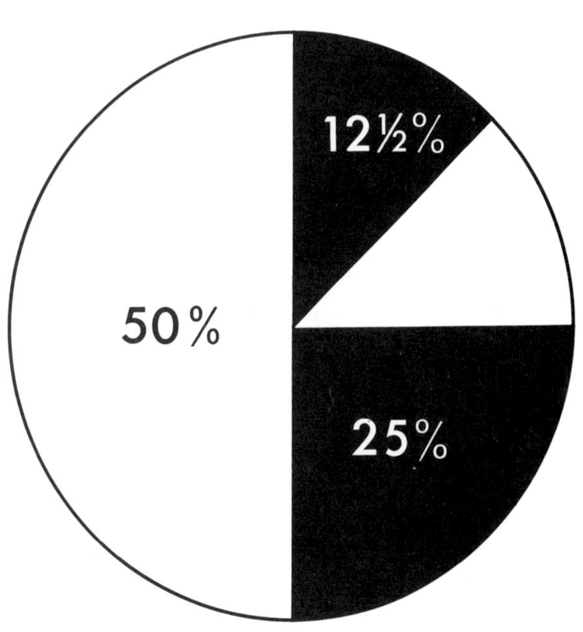


Pie graph

## Drawing large circles

Sometimes it is necessary to draw a circle or part circle that is larger than the normal compass span will allow. There are several methods which can be employed to give reasonably accurate results. For instance, a pin placed through the end of a 60 cm (2 ft) steel ruler and a pencil held up against the desired measure along the ruler's edge will give a reasonably satisfactory curve, providing pressure is kept even whilst moving the implement through its arc. Another method is by means of a piece of wood to which can easily be fixed a centre pin and pen or pencil at the other end by means of wrapping around tightly with *Sellotape* to secure to the beam. Another method is to use a cord or string that has very little stretch pinned to the centre of the circle, the pen or pencil being held firmly at the desired radius. A cheap form of beam compass that can be readily and easily made by the average handyman is shown.

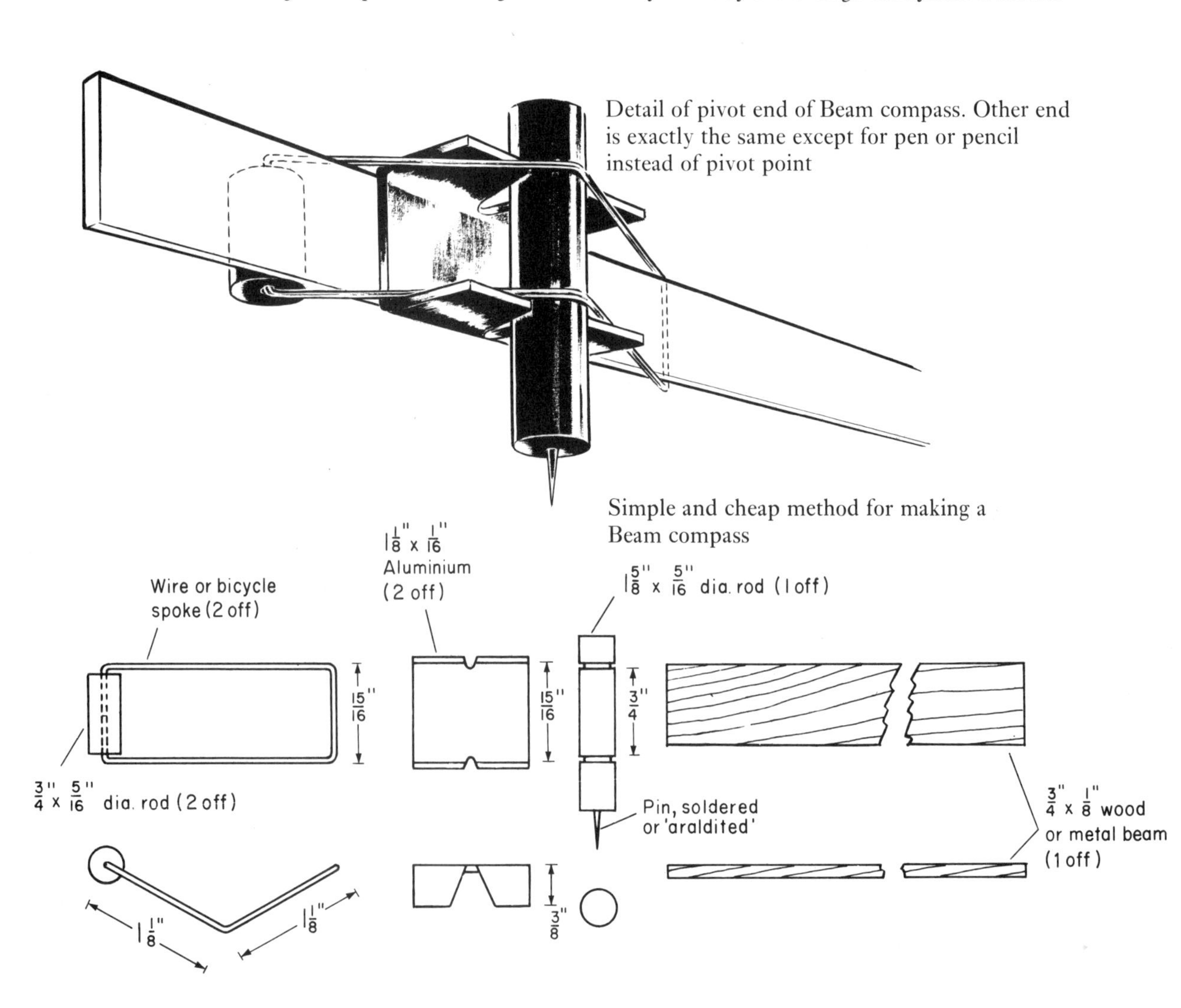


Detail of pivot end of Beam compass. Other end is exactly the same except for pen or pencil instead of pivot point

Simple and cheap method for making a Beam compass

### Stretching paper to take colour washes or body colour

Often it is necessary to stretch paper in order to be able to work satisfactorily upon it, particularly where flatness is essential for accuracy and ease. Where the colour to be put down is water colour or body colour, the paper should first be soaked, preferably in a sink of water (failing that, it should be given a heavy sponging), for approximately five minutes. Then take the sheet of paper out of the water and place it on a clean surface, *not wood*, dab off the excess water with either blotting paper or a soft cloth such as a tea towel. Place the paper on a suitable surface, such as a sheet of plate glass 9 mm ($\frac{3}{8}$ in.) thick, *Perspex* (*Plexiglas*) or mounted *Formica*, with the right side of the paper facing upwards. Then take four pieces of gum strip paper, approximately 50 mm (2 in.) longer than each of the edges it is to cover, stick down the paper on to a backing board, and if the whole sheet of paper needs painting, apply the colour while it is still in its wet state and allow it to dry naturally. Do not attempt to dry with direct heat, otherwise if the sheet of paper is large, it will probably dry unevenly and tear the gum strip paper. If speedy drying is essential, then artificial warmth should be used at a distance, preferably from a fan blower. It will be found that once the paper has dried, one can afford to be fairly lavish with the use of water or paint without the paper cockling and forming pools of water or colour.

*Warning*: certain woods when wet produce a permanent stain on paper and therefore should not be used. I would suggest that if a stretcher board, such as those I have mentioned, is not available and a wooden one has to be used, it is covered first with a sheet of cheap paper (this too should be soaked) that can be disposed of after use.

Certain colours can be worked for long periods on paper prepared in this way without undue damage to the surface of the paper. The least amount of colour should be used when using body colour, as there is a tendency when the paper is removed from the stretcher board for some colours to crack through lack of adhesion to the paper surface. This is particularly prevalent among the bluish-reds and mauves. Yellow is a particularly good colour and is generally the easiest colour to apply as it lays flat. A large brush should always be used wherever possible and the colour brushed from side to side and end to end without stopping midway, otherwise marks will ensue.

A useful tool in the studio for laying large areas of flat colour is a sponge type rubber roller such as is used in house decoration. Perfect results can be obtained by this means.

### Dividing a line of any length into equal portions

This is a useful piece of schooldays geometry that may be required at some time or other. The line to be divided may not be of any measurable length and the way to divide it is as follows: from one end of the line at an angle of approximately 30–45° draw another line and divide this up by means of a ruler into the desired number of lines or spaces. Having marked off the line, take a ruler, set-square and a pencil, place the set-square on the edge of the ruler and connect the furthermost points of both lines together. Then run the set-square along the edge of the ruler marking off as one goes.

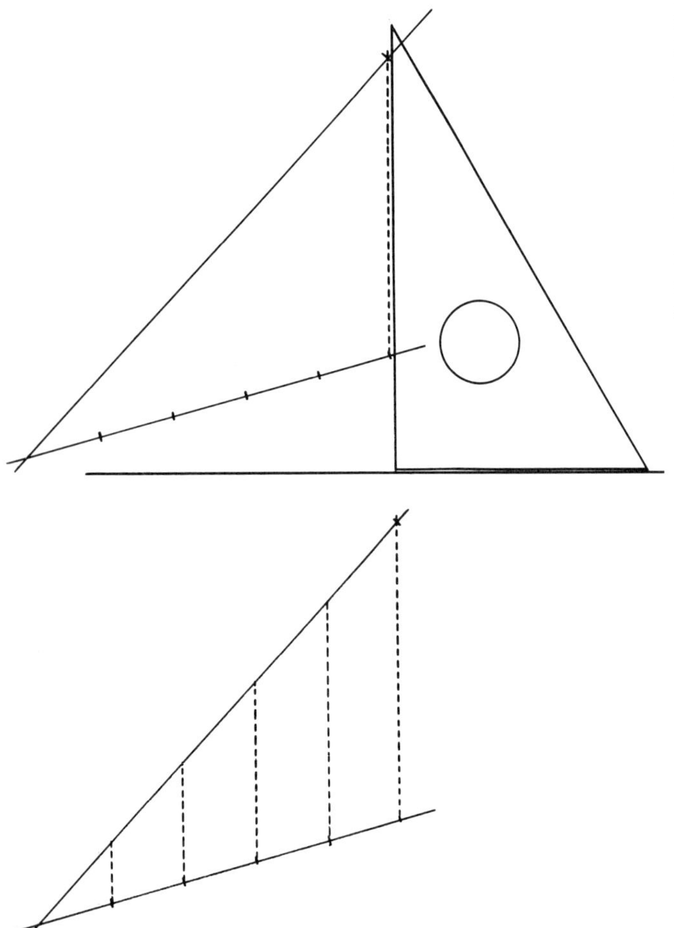

### Widening existing artwork

This is a problem that sometimes arises, particularly where a piece of master artwork needs to be altered to fit a larger space. Generally speaking, to save completely re-drawing such things as fancy borders, the paper may be stripped from its backing board: take a sharp blade and carefully cut the paper surface only, then peel back one half of the paper and completely strip it from its backing. Some boards are readily strippable whereas some may prove troublesome and tend to tear unevenly. Should this be the case it is useful to have a cardboard tube nearby that the paper may be rolled around as it is peeled from the backing card. A piece of similar quality paper may then be rubber gummed and inserted to the required width, so, in effect, there is a paper gusset. Extra card may need to be added on the outside edge to support any overhang of paper. Where this is done it is recommended that the whole of the piece of work is then mounted on to another board to act as a support. The same technique may be applied of course for shrinking work, merely by lifting the required parallel strip from a convenient point and moving the remaining pieces of paper closer together. Never dispose of portions of artwork that are removed – they may be required again. Joints should be as accurate and tight as possible in all cases to ensure that when the work is put in front of the photographer's or the blockmaker's cameras, no unnecessary lines or shadows occur.

Dividing a line of any length into equal portions

### Overlays for stripping guides

On some types of job it is necessary to use an overlay drawing for use by the blockmaker in accurately positioning an extra working. For example, lettering may need to be put down precisely in position on a tone photograph as a solid line portion without the half-tone screen cutting up that lettering; conversely the lettering may need reversing out of the halftone so that the letters come out as white. The usual method employed is for the master artwork to be on board with a *Kodatrace* overlay that is *Sellotaped* along the top edge and all lettering or repro type setting is mounted on this overlay by means of rubber gum. The usual practice is then to draw two or more register marks (see illustration page 36) on the master artwork and the same number of register marks on the overlay or cut window holes in the overlay. The blockmaker will then take the halftone portion of the artwork and photograph it through his screens; he will do likewise with the overlay portion of the work but make a line negative which he will print down upon the half tone portion, using the register marks as his guide. This is commonly known as 'stripping in', but may be known as 'double print down' or DPD.

Overlays for the *second* or *third* colours are treated in much the same way and can either be drawn as 'keylines' or as solids, again using the register mark technique to help the blockmaker. Either indian ink or paint may be used on these overlays without much fear of them cockling. I would stress again that it is most important to read manufacturers' instructions regarding use of certain materials, that rubber gum on materials such as *Ethulon* must not be used under any circumstances, as this causes distortion of the material. The usual practice for second or third colours is to paint them in solid, and where the colour is to meet or surround the master artwork, simply paint up to it accurately. Alternatively, a keyline may be employed on the extremities of the overlay and offsetting can be done by the platemaker which can prove to be rather expensive; it is far better to do as much hand work on the original as possible because, generally speaking, it is not only quicker and cheaper but more accurate.

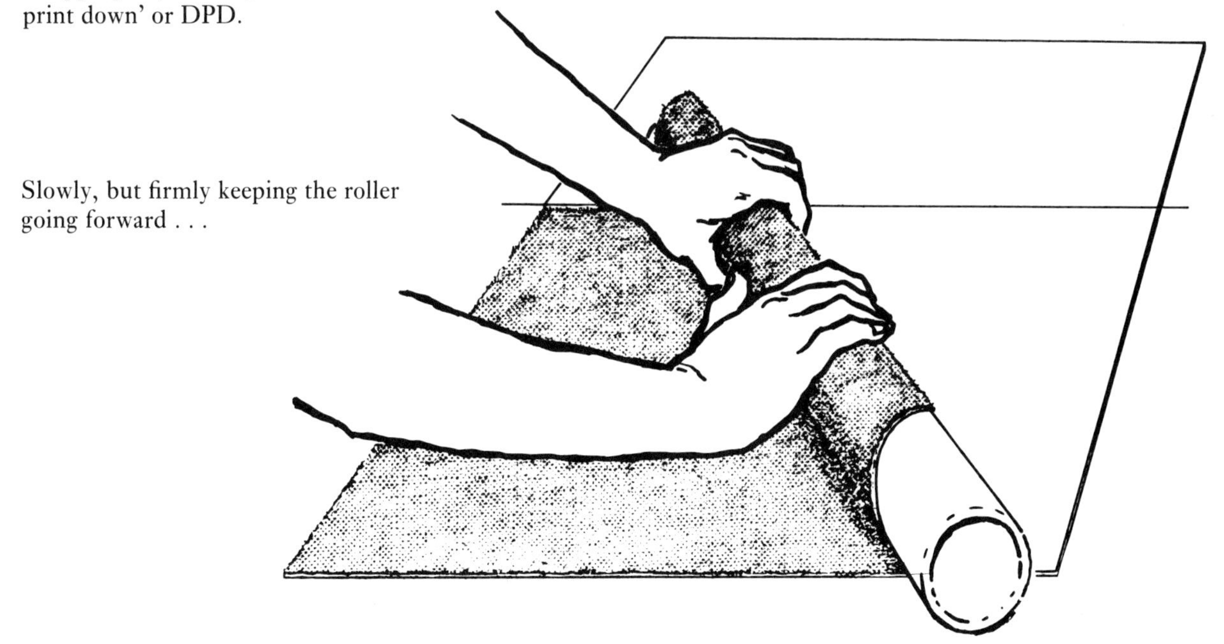

Slowly, but firmly keeping the roller going forward . . .

## Scaling up and down

It is usual practice when preparing artwork for reproduction (particularly where the final product is going to be small or parts that are difficult to work on), to work up in size and such terms as 'quarter up', 'half up', 'twice up', are commonly used. In other words, the finished artwork size is done to a convenient but proportional size to the printed size.

Assuming that one has an accurate rough from which to work, there are five ways in which an accurate finished artwork may be obtained.

1 By scaling up by use of diagonal lines. This is extremely laborious and slow, and has tremendous drawbacks when dealing with a job that is rather complex in shape.

2 By means of proportional dividers which have an adjustable thumb screw sliding in a groove, and a set of points at either end. These can be bought in varying sizes and used with great accuracy on any job that requires enlarging or reducing. They are expensive items but soon pay for themselves. The short ended points are placed across the object to be enlarged and the large points at the other end are the size that the object will be enlarged to, according to the scale that has been set by the thumb screw.

3 By *Pantograph* which is a simple adjustable wooden frame that can be set trace by an indirect method. I have never seen this used in a commercial studio as it tends to be rather unstable and slow in use.

Lines bisecting A–C–B and AA–CC–BB converge on R. All other lines viz W–X and Y–Z are parallel

Diagonal method of scaling up or down

'R' B BB A C AA CC W X Y Z

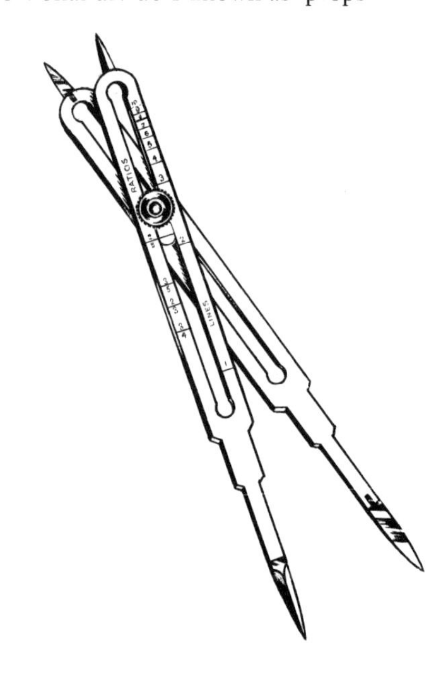

Proportional dividers known as 'props'

4 This is also rather slow. First of all draw a series of squares on a piece of flimsy paper laid over the rough and then draw the same number of squares but in an enlarged form on another piece of tracing paper and laboriously copy line for line.

5 The quickest method is with the projector.

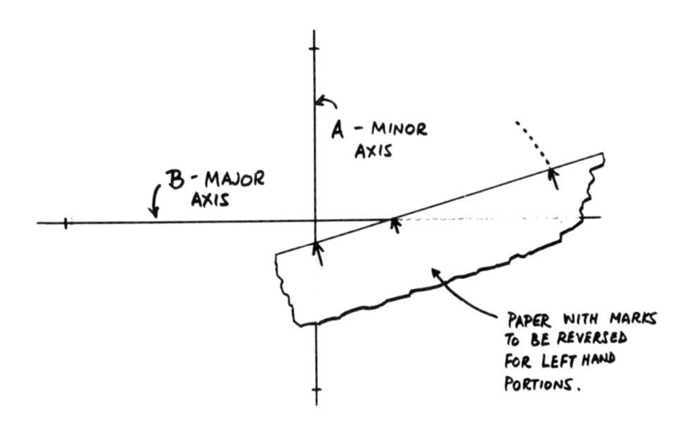


## Ellipses

Generally speaking there are three methods which may be employed.

1 This is the old method of drawing a major and a minor axis at right angles to each other, determining the curve of one quarter by eye, then drawing this as accurately as possible, and then tracing it off successively to fit around the other three quarters. This is a hit and miss method and one usually finds that it either tends to end up rather pointed on the major axis or rather too full. It is particularly difficult when an ellipse of reasonable size is required. It is advisable to use this method only where the major and minor axes are no more than 76 mm × 50 mm (3 in. × 2 in.).

2 This is probably the best of all and is again to draw up a major and minor axis at right angles to each other, and then either by use of a ruler or ticking off half the major and half the minor axis on a straight edged piece of paper, make a series of dots trammelling one quarter of the ellipse at a time. The more dots that are made, the more accurate is the outline which can be finally joined by hand, either with pen or pencil.

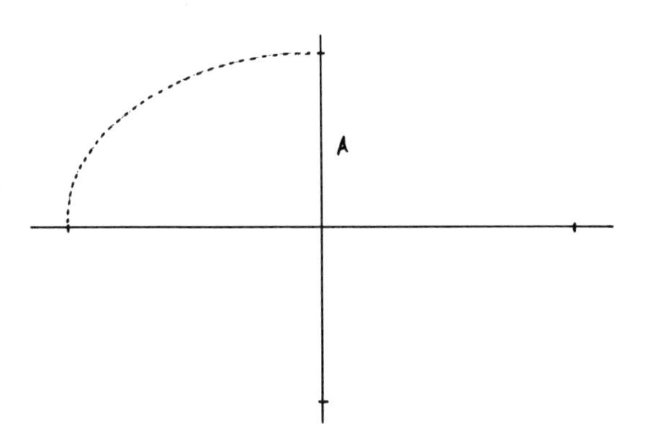


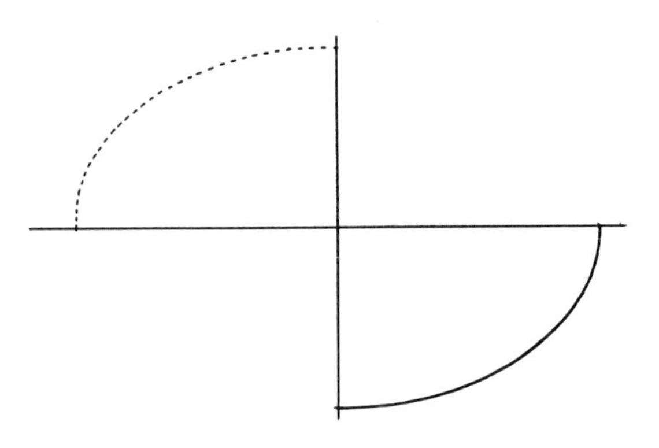

3 This method is a patented device known as a *Rulalipse*. This has four disadvantages:
(i) It will draw only half an ellipse at a time;
(ii) Fixing pins have to be used, which can result in rather unsightly holes in the board or paper;
(iii) It is rather limited in size;
(iv) When reversed to draw the other portion of the ellipse, difficulty will be met in making both halves meet exactly.
Briefly, the method by which this works is an adjustable swivel arm that will take the point of either pen or pencil and a bar of metal moving in a vertical channel. An adjustable arm is able to move in a circular motion and the vertical arm is able to move in one direction only. The lower adjustable screw as shown in the illustration, is used in conjunction with the outer screw to set the machine up for major and minor axes.

## Dry brushing and pen stippling

*Dry brush* work is used a great deal in advertisements that are being printed on newsprint. The most difficult thing to achieve is the correct consistency of body colour, and to be able to maintain this consistency. It is easier to work on a coarse tooth paper, such as *Rough Kent* or on a water colour paper. Generally, dry brush work is not easily done on smooth hard surfaced types of paper.

*Pen stipples* are not very often used today, as it is a slow and laborious means of tinting drawings to give them depth and shadow, but can be used on small areas that would not justify the cost of laying a tint. They were quite commonly used 15 to 20 years ago when time was of less consequence.

Dry brushing

Rulalipse

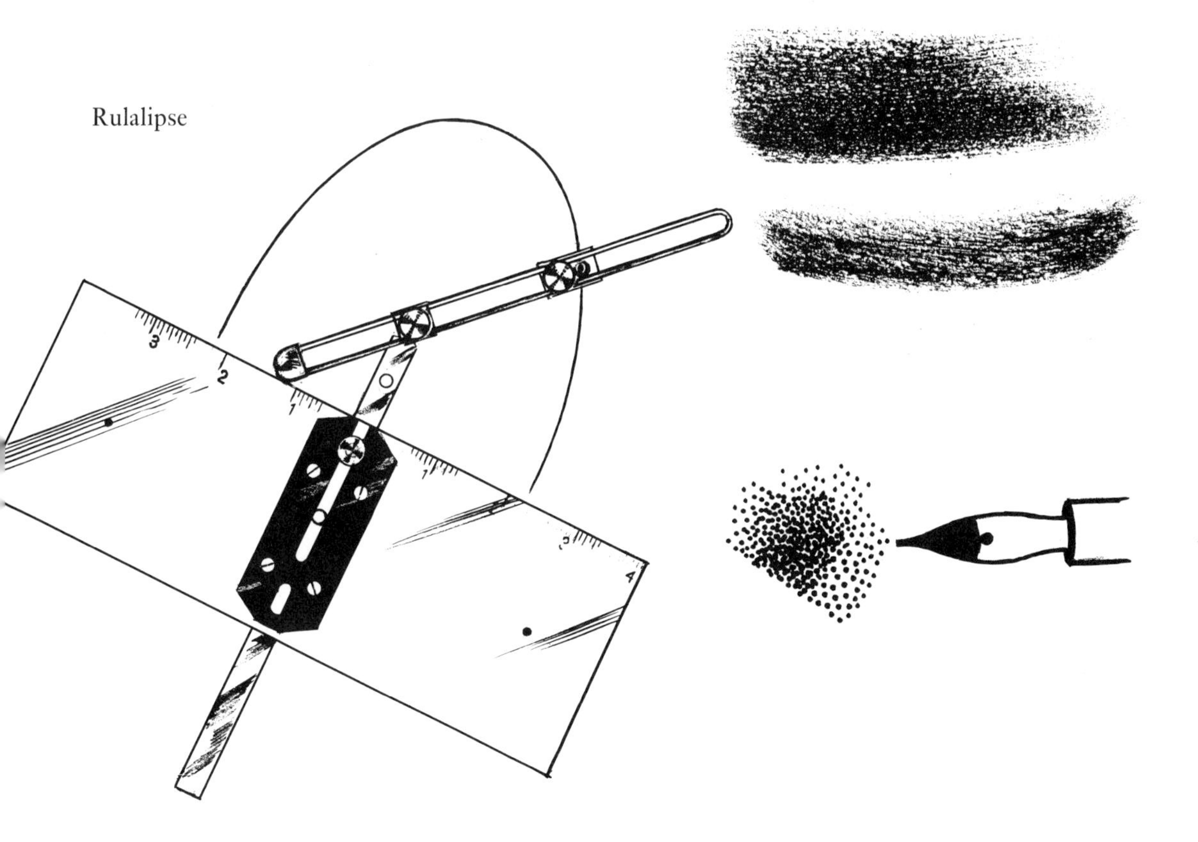

## Erasing mistakes

Depending on the medium used and the surface on which it has been used, mistakes can sometimes be troublesome to correct. It is as well to have the following items to hand to cover any eventuality – safety razor blades, glass fibre pencil, sand paper block, distilled water, blotting paper, cotton wool, *Milton* sterilising agent used as a bleach. The razor blade is a favourite means of removing either paint or indian ink from paper or synthetic surfaces such as *Kodatrace*. Certain colours are easily removable, whereas others can be rather troublesome. For instance, designers' or water colour lamp black has a tendency to sink well into the paper, whereas designers' jet black can be almost completely erased by washing out with clear water. Excess colour can be mopped off with blotting paper and then when perfectly dry, finally erased with a good sharp razor blade. Sometimes it is best to use a one direction movement of the razor blade, so that the surface does not become too coarsened. Glass fibre pencils may be used in much the same way, but care should be taken as there is a tendency for them to tear the surface of the paper rather badly should one become at all heavy handed. They also have the disadvantage of being rather too local in use.

A fine sandpaper block, such as flour paper, may be used in certain instances but it is as well to do a test strip before attempting to erase anything by this method, as colour tends to get scrubbed into the surface and the result is a messy looking piece of paper.

Distilled water is particularly useful where splashes of indian ink have occurred, either on the art work or on one's clothing. This can be readily obtained at the local garage or chemist.

*Milton* can be used as a final cleansing agent on certain jobs and can be applied on a wad of cotton wool in a neat state, and gently washed across the surface, keeping it as local to the mistake as possible. It is as well to try all of these methods thoroughly as an exercise, so that in the eventuality of mistakes occurring, one is prepared.

Even household bleach can be used with some success on certain jobs.

## Delicate paste-up work

Cutting and mounting individual letters or numbers can be a problem where the letters or numerals are to be mounted down square to each other, and it is not always practical to use a ruler and set square or T-square. Therefore a simple guide is needed to get round this difficulty. Supposing one is preparing artwork of a street map and it needs letters or numerals mounting on an overlay; to make sure that the letters or numerals are square, first of all draw up a grid on flimsy paper. Place this grid of approximately 25 mm (1 in.) squares underneath the *Kodatrace* overlay then take the sheet of letters or numerals, apply two-way tape to the back, then very carefully cut parallel, horizontal and vertical lines in between the lines of numbers. Pull off the protective backing from the two-way tape then by use of a divider point or scalpel blade, place them lightly in position on the overlay and when satisfied that they are square, press them firmly down, using the pencil guide as a visual aid. Where a *Kodatrace* overlay is not used, then it may be found necessary, if there is no grid drawn on the map, to lightly draw one in pencil on the artwork or to turn to the aid of a ruler and set square or T-square and set square.

## Repetitive ruling up jobs

Some ruling up jobs such as large office diaries that are divided up into mornings, lunchtime and afternoon sections as well as days, can be rapidly dealt with with a little thought and practice. First, draw up the page areas and then mask vertically with *Sellotape* where the lines are to begin and end (see illustration). This masking medium enables one to rule across two or more pages without lifting the ruling pen from the paper, drawing over the masking tape to the very end of the paper. When the ink or paint has dried, the masking medium is simply stripped off leaving clear white gutters. Also, if the day and date have to be mounted down in position, this can be done immediately the ruling has been completed, again working across the pages two or more at a time.

## Scraper board (sometimes known as scratch board)

This is heavily coated with china clay and can be obtained in white or black. The type of scraper board illustrations seen in magazines and newspapers are generally of a highly mechanical and detailed nature, worked from tone photographs as reference and even in some cases used as a tracing. Scraper board was a particularly popular method in the years 1940–1952, this being due to the shortage of good newsprint. It has the advantage of being able to reproduce on any surface of paper with utmost clarity. It is also one of the cheaper methods of processing, as it is produced in zinc, but from the illustrator's point of view it is an extremely exacting and expensive method of working. There is also a method which some processing houses are able to offer which is similar to scraper board but is based on a continuous spiral line.

The illustrations show the various thicknesses of line to obtain light and shade.

## Masking media for spraying

There are several methods that can be used for masking areas which are not to be sprayed with paint. Probably one of the most used is *Sellotape* or *Scotch Tape*. This is suitable for very small areas, depending on its width, or for sticking down the edges of large sheets of paper to protect big areas. Where delicate work is encountered, paper known as *Frisket* or *Glassine* is stuck to the background with rubber gum and peeled off readily when the spraying has been finished, leaving a thin coat of rubber gum to be removed as described on page 24.

Another masking medium, particularly for intricate work, is photographic masking medium, that is a bright pink in colour and is painted on by brush and allowed to dry. When the spraying has been completed, this is peeled up in strips by means of a small tab of *Sellotape*, leaving the protected area clean and clear.

A clear adhesive film, marketed under the name *Transpaseal*, can also be used most successfully and, if carefully removed, can be re-used several times on different jobs.

A very old method that is never used now, but is interesting to note is that of using the white of an egg mixed with process white, which can be chipped away when dry.

## Dummy boxes

These are sometimes required for use in packaging design and are best made with carton board which can be obtained from art board manufacturers. It is very hard, thin and flexible, with a working surface which is white and fairly hard. The underside is a light straw colour. Easily creased and folded, it takes body colour reasonably well. When scoring for folding, it is best to use a ruler and the back of a knife or similar blunt instrument to bruise the surface of the carton board, thereby allowing it to bend easily. It is simple to cut and work, and can be shaped most readily. Tent cards can be made from the same material, providing they are small, as there is insufficient stiffness in the material for them to be very large.

## Mounting

The mounting of such things as photoprints for display or retouching purposes falls into three sections: wet, dry and rubber gum.

*Wet mounting* as the term implies, is where the photoprint has been immersed in water and excess moisture has been wiped off. Then a coat of paste such as one uses in paper hanging or of a latex variety, has been applied to the back of the photograph and possibly to the surface on which it is to adhere. If the photoprint is not wetted to begin with, it will generally cockle rather badly and be very difficult to lie flat, but through soaking in the tank of water, the stretch becomes very even. It should be remembered, however, that when a photoprint is immersed in water it will expand to quite a large degree *viz* a print 38 cm × 30 cm (15 in. × 12 in.) can expand 15 mm (½ in.) or more along one dimension. A print of 137 cm (54 in.) width will certainly expand up to 38 mm (1½ in.). Of course, as the moisture departs from the paste and the photoprint a shrinking effect will take place: with the shrinking a 'bowing' of the board will occur, so much so that a board of 0.050 calibre with a 38 cm × 30 cm (15 in. × 12 in.) print mounted upon it will probably bow 76 mm to 102 mm (3 in. to 4 in.) or more. Obviously then some step must be taken to prevent this bowing or dishing action. The first thing that comes to mind is, of course, to use a heavier weight board such as calibre 0.135 (3 mm (⅛ in.)). This will prove quite satisfactory on small photoprints, but is still likely to bow on prints above whole plate size (20 cm × 15 cm: 8 in. × 6 in.). If, however, for some reason, a thick mounting card cannot be used, then the answer must lie in using a backing paper to pull the board in the opposite direction so that the photoprint and the backing paper have a cancelling effect on each other, thereby enabling the board to remain flat. Care must be taken in ensuring that the grain of the photoprint and the grain of the backing paper are running in the same directions. To ascertain this, before either piece of paper is immersed in water, a measurement should be taken on two sides. When both pieces of paper have been immersed for the same length of time they should be measured again. It will be found that there is a positive expansion in one direction and very little in the other. Obviously, both pieces of paper should have their maximum dimensions running the same way, not at right angles to each other. Where it is intended to use hardboard with ordinary paper paste, the hardboard should be 'sized' first of all on the rough side. This will prevent the paste's moisture being sucked out before the print has had time to set. It is preferable to use a ready sealed hardboard, as this is less prone to stain wet photoprints, should the image side come into contact with the hardboard. Where there is a large overhang of photographic paper when the print has been mounted, trim off the excess flush to the edge of the board, with a sharp blade and in the case of hardboard even a wood working plane can be used satisfactorily when dry, finishing off with a sandpaper block to remove paper burr. Where large photoprints are to be handled such as those for exhibition work, it is advisable to have a squeegee close at hand and, if possible, a second pair of hands is of great use. When both surfaces have been pasted, on bringing together, work from the shortest edge and run the squeegee right down the middle of the photoprint and then work out sideways to exclude all air. This should be done gently, or an irreparable crease may occur. If a squeegee is not available, a damp cloth, such as a tea towel bunched up firmly in the hand, will do well. Wipe all excess paste from the surface and edges and allow to dry quite naturally, or with a cool fan blowing gently over the surface, not allowing any air to push up the edges.

*Dry mounting* is a reasonably simple operation that can be carried out either with a proper dry mounting machine or with an ordinary electric iron. Here, instead of the medium of paste, we have the tissue that comes in roll form and is a shellac type of material that, when attacked with heat, goes soft and adheres to paper or board. It is a medium that is readily used for mounting photographs for retouching or for exhibition purposes; the main advantage being that dry mounting is almost instantaneous, there is no waiting time once the picture has been put in position. It can be worked on immediately by retouching or dyeing. As opposed to wet mounting, dry mounting curves the opposite way, this being due to the mounting board being thicker than the photoprint; it contains more moisture and when going through this heating process, a certain amount of moisture is withdrawn from the card, thus causing shrinkage. Hardboard may be mounted upon either surface satisfactorily, but the patterned side will leave a distinct mark. Where very large areas are to be dry mounted, it is often the practice to move the photoprint and mounting card several times in piecemeal fashion in order to cover the whole of the area.

*Rubber gum* (cement) is the most commonly used of all the mounting media in the average studio. It is quick and clean, but has one disadvantage in that its use is of a temporary nature only. By temporary is meant a period of months, or possibly weeks, before drying out takes place and edges begin to lift and curl back. The larger the area that can be covered with the rubber gum at one stroke, and the thinner it can be applied, the better it will work. Both surfaces to come together should be dry to the touch and, again, as far as possible, working from the shortest edge lay the photoprint gently on the area and press outwards with a clean dry cloth or cotton wool to exclude all air. Where a print is prone to cockle at the edges, it can be nicked with a sharp knife to get it to lie flat, if it happens to be an unwanted area. If the whole of the photoprint is needed and prone to cockling, it will be found best to damp the back of the photoprint first of all, or even immerse it in a basin of water for a few minutes to take the unevenness out of the surface. Sponge off all excess water and then rubber gum in the normal way. It should be noted that whatever method is used for mounting photoprints, where they are to be retouched, the surface should be thoroughly cleaned with either clear water, high grade petrol, surgical spirit, methylated spirit or last, but not least, ordinary spit. This must be done particularly if the area is to be air brushed, as any finger marks that are left on a photographic surface will show through and cannot be covered by airbrush spray.

Two-way tape is also a useful medium for mounting small prints up to 76 mm × 102 mm (3 in. × 4 in.). It is not practical to go above this size due to cost.

Wet mounting

Dry mounting

# Photographic Techniques

## Copy negative and bromide prints

One of the most useful services a studio can find is that of a 'Copy Neg and Print Service', as it is generally known. These are studios that specialise in copying either artwork or photographs and printing them out on bromide paper. Process plates are used for line work such as copies of lettering or line drawings, and continuous-tone plates used for copying any tonal work. Where colour work is to be copied but transmitted in terms of black and white, either panchromatic or orthochromatic plates are used, because they are more sensitive to certain colours of the spectrum, and will either reproduce as dark greys or even black, or not pick up other colours at all. Of course, where an original tone picture is to be copied, it will invariably lose a certain amount of quality when copied. Wherever possible original negatives should be used. Where colour transparencies are to be copied and transmitted into black and white, the photographer will place the transparency upon a light box with a suitably cut opaque mask, and then copy in the normal way. There is a tendency for prints from the ensuing negative to be rather 'hard' and, generally speaking, inferior to those that are copied from black and white works.

## Bromfoil prints

For this particular type of print an ordinary negative is made and a normal printing out action is taken. The particular use of this type of print is that, provided the photographer is absolutely accurate in his measurement on the sizing up board in the dark room extreme accuracy is guaranteed, due to the fact that in between the layers of paper, one of which is sensitised, is a thin layer of metal foil which is bonded in. This type of print is particularly useful for colour separation work where utmost accuracy is essential. By means of making two or more negatives from an original drawing and then blocking out unwanted parts on each negative, it is possible to end up with colour separated drawings in a photographic form that are one hundred per cent accurate. The disadvantage of these prints is that they come in a matt surface and do not reproduce good mid-tones. A further disadvantage is, if one wishes to move a portion of a drawing or lettering, it is extremely difficult to cut this type of print. Price wise, they are something in the region of 2–2½ times as expensive as normal bromide prints.

## Autone prints

Here again a normal copy negative is made and printed down on to specially sensitised paper that has a built-in colour base. The range of colours is not very great but extremely useful. Included in this range of colours are silver and gold, which are useful to exhibition designers for display purposes. In the course of processing it is possible for the same photoprint to be either a coloured background with black lettering or image, or a coloured background with white lettering or image. This is due to the fact that through processing, the emulsion is soft and may easily be wiped off with the finger tips or cotton wool whilst it is in this state. They are suitable for both line and tone work.

## Retouching photoprints

This is a very necessary and important part in commercial art. So much so that there are artists who specialise in particular sub-divisions of the work. For instance, 'knife work', i.e. use of the scalpel, applies particularly to figure retouching, giving a positive and yet direct method of working, and enabling such things as black spots or hair lines to be completely erased, or even to reshape parts of a face or figure. It is only when turned at an angle to the light that this method of work can be detected. It requires a great deal of skill and patience besides an extremely sharp blade.

Chemical retouching is done with varying strengths of bleaching solution at controlled temperatures. When skilfully done, no trace of retouching is apparent.

Airbrush retouching is used a great deal in the mechanical field, such as large shiny areas of car door panels, bonnets, etc. It is particularly useful where graduated tones are required and if it is fine screen production. Only under the most dire circumstances should it ever be used on figure retouching, particularly faces, as it tends to give a rather waxy or metallic look to the subject, producing an unnatural effect.

A certain amount of brush work such as stippling or spotting is required on most photoprints. Contrary to belief, the colours required to make good matching greys are not only lamp black and process white, but also water colour sepia, which gives the correct depth of colour and warmth. It is, of course, possible to buy sets of greys, specially numbered 1 to 5, but for the ordinary everyday artist, who is confronted with the occasional piece of retouching, I would suggest that these are a waste of money, as they tend to dry out in the tube rather quickly. Therefore it is better to make up one's own greys as and when necessary. When applying paint to photographic surfaces, care should be taken to avoid the build up of any lumpy mass, or even when photographed through the coarsest screen, there will be a likelihood of it picking up and showing on the printed image.

Colour transparencies are often retouched but this is a highly technical procedure and should the reader require information on this special subject, process houses will usually readily give any information and help. Failing that there are several very good books on the market.

## Cut-out tones

This is the term used for photographs that are irregular in shape, instead of the normal squared up look. Generally speaking, the photoprint is mounted on board with rubber gum and then the desired portion is whited round very carefully with process white for about 6 mm–12 mm ($\frac{1}{4}$ in.–$\frac{1}{2}$ in.). This enables the block-maker or platemaker to see quite clearly on his negatives where he is to paint out unwanted screen.

## Montage of photoprints

Occasionally one is called upon to do some faking in the form of montage. That is, bringing two photoprints together in such a way that they appear to have been originally photographed as one. For example, it may be necessary to cut out around the heads of people in a picture of a crowd scene, and move them elsewhere in the picture. This is best done by having two photoprints of the same size and then by laying a roughly cut area of one photoprint over the other, ascertaining where finally it is to be put down. Then mount the photoprint with rubber gum on its intended position and cut carefully around the wanted portion with an extremely sharp scalpel blade, giving sufficient pressure to it to cut through the photoprint underneath at the same time. Peel off the montaged part of the photoprint and then take out the portion that has been cut on the background photoprint and insert the montaged photoprint in its place. This should, in effect, come as near a perfect fit as is possible if care is taken. Should, however, the knife slip and the edge of either print become damaged, it will mean that the montage is no longer clean and accurate; it will then be necessary to clean the rubber gum off the back of the top photoprint and carefully black all round the cut edge with paint or ink. Allow to dry for a minute or so, re-rubber gum and stick back in position.

Where this method is not suitable, the following may be done: presuming this to be a fairly complex shape, turn the photoprint over with the emulsion side down and carefully go around all the edge with an almost flat, sharp scalpel blade and feather the edge off as if one were carving at meat, finally finishing off with very fine flour paper (which is a very fine form of glass paper), so that the edge of the paper is as thin as possible without tearing. Then rubber gum both surfaces to come together, allow to dry and then bring together, burnishing down the edge well with a finger nail. Remove all excess rubber gum by working away from the mounted photoprint, either with the finger tips or a bungie. Do not, under any circumstances, work towards the cut edge, otherwise the photoprint will lift and tear. This is a particularly useful method where heads have been cut out and it is wished to put back little wisps of hair to give a natural effect, softening off and merging with the background.

## Opaquing negatives

This is an uncommon job to find its way into the commercial art studio, but nevertheless it is important to understand and to know what is desirable. Wherever possible, be generous with the amount of paper left around the image so that when it is photographed and probably reduced in size to fit the negative, the cut lines that are picked up by the camera are not so close to the image as to make opaquing a long and tedious job. This should be borne in mind, particularly when an urgent job is sent to the photographic studio with the message 'please may we have this back in an hour?', otherwise the quality of the job will inevitably suffer. Some negatives such as copies of tones are sometimes required to be 'blocked', either geometrically or irregularly, depending on the subject; for instance, one may have a negative of a piece of machinery, e.g. a typewriter, which has been photographed on a normal office desk with perhaps one or two other pieces of office equipment beside it. It is then decided that only the typewriter is wanted and on a perfectly white background. This is where the opaquing medium comes into use. Generally, it is applied on the emulsion side of the negative as it tends to bind better. Where two negatives are to be bound together it may be necessary to block one negative on the non-emulsion side to prevent them sticking together. The binding of two negatives together is used where the best of both worlds is sought, i.e. good tone and good line-work. It must be understood that a straight-forward copy negative, if it is shot on tone film, will only give good results of the tonal portion and a rather grey muddy effect in line work that is on the same piece of artwork. On the other hand, if a line negative is used, the line portion will be good, but the tonal portion will lose a good deal of its middle tone range and become very hard. So, in instances such as this, it may be possible to make two negatives from the same artwork, cut out the respective portions and bind them together by means of clear tape and the use of opaque on portions not wanted.

Using the light box for opaquing a negative. Note black mask to reduce amount of light being thrown out

## Bleach outs

Often it is necessary for work of a highly detailed nature to be photographed first of all and converted from halftone into line. The easiest and best way to do this is to have a half strength photoprint made from the negative and printed on a matt surface paper (in very detailed work it is a wise plan to have a good quality full strength print on glossy unglazed paper to act as a reference). On receipt of a matt print the first thing to do is to mount it on a mounting board with rubber gum. This will give a firm background on which to work. The next stage is to draw in all wanted detail in fixed indian ink with either pen or brush. When the desired amount of work has been done, gently but firmly remove the print from the mounting board (should this prove troublesome, soak with petrol and allow to dry) and clean off the rubber gum from the back. The whole of this may now be immersed for a few minutes in a weak solution of potassium ferro-cyanide and photographic fixer. This bleaches out the remaining photographic image from the paper, leaving the pen and ink lines that are needed. Withdraw the pen and ink drawing from the chemical bath and wash it gently in running water for a minute or two, then dab dry with blotting paper, or allow to dry in its own time. When dry or almost dry, the drawing can be remounted on mounting board and is then

Full strength bromide

Half strength matt print

ready for the final stage in working up, with either paint or pen and ink. This type of illustration is particularly useful on exploded type drawings that are found in technical books. For special effects it is possible instead of bleaching out chemically, to make a series of negatives from an original halftone negative, and printing them down on to very hard paper, gradually losing all the light tones and ending up with only really solid portions.

Partly bleached line drawing

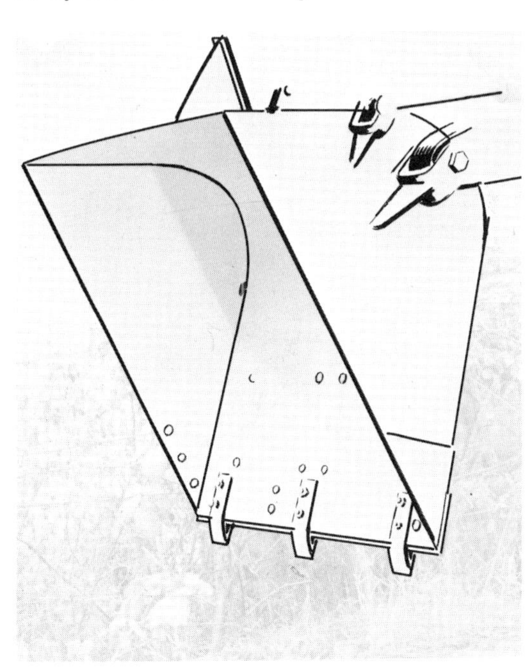

*Potassium ferricyanide and hypo* (Farmer's reducer)
A yellow solution used as a general reducer for lightening normal prints or completely bleaching pale grey prints on matt paper for 'bleached out line drawings'.

Carefully wash to remove stain and also to avoid possible damage to ink work on surface.

*Iodine-iodine reducer*
A mauve coloured solution that can be kept and re-used, ideal for bleaching out. Gives a very clean result on all densities of photographic image.

The print absorbs the iodine solution during process and becomes stained deep mauve. When the photographic image is no longer visible, the print is cleared by placing in a normal hypo solution until all colour has disappeared.
Wash carefully.

*Iodine-iodine-cyanide reducer*
A very poisonous clear solution. It should be handled with great care. The solution should never be used hot or whilst the handler is smoking as fumes that are inhaled are extremely poisonous. Neither should it be used if cuts or grazes can come in contact with the solution.

The process of bleaching can be easily observed, controlled or stopped at any stage by immersion of the print in hypo solution. Wash carefully.

With all the solutions mentioned – the stronger they are, the quicker they work.

Again, take warning – all these solutions are poisonous. After use carefully wash hands and utensils thoroughly.

## Formulae

*Farmer's*
Make a 10% solution of hypo with 10% solution potassium ferricyanide. Add a few drops of the ferricyanide to some of the hypo in a dish to make a weak solution.

More 'ferri' can be added to speed up the process, also to replenish the reducer during use as it quickly deteriorates.

*Iodine-iodide*
Place a level teaspoonful of potassium iodide crystals into a small bottle together with about one quarter that amount of iodine flakes. Add a very small amount of water (say 1½ teaspoonsful) and shake the bottle. When dissolved, more water can be added to fill the bottle (say, ½ pint – 0.284 litres). This is a strong working solution that can be used repeatedly. Take care not to contaminate it with hypo solution.

*Iodine-iodide cyanide solution*
Make up a small bottle of 10% potassium cyanide solution. Take some of the working solution of the iodine-iodide reducer and add a little cyanide solution until it becomes colourless, then add about the same amount again. Now ready for use, more water can be added for slowing up the process if necessary.

*Bleaching*

| | | |
|---|---|---|
| (i) Potassium iodide | 28 g | 1 oz |
| Iodine | 14 g | ½ oz |
| Water | 567 g | 20 oz |
| (ii) Potassium cyanide | 28 g | 1 oz |
| Water | 567 g | 20 oz |

Ten drops of solution (i) and (ii) to 283 g (10 oz) of water. Pour on to print, leave for one minute and allow to clear, wash in water.

## Glazed prints

These are photographic prints that have been put round the drum drier with the emulsion side facing the highly polished metal drum. Consequently, the surface, when dry is mirror like and very hard. This is not suitable generally speaking, for working on either with brush, airbrush or knife. Therefore it is necessary to remove this glaze, which is done simply by placing the print in a basin of cold water and leaving it for 10 minutes or so to soak, then drying off excess moisture by dabbing with a clean cloth or by means of a drum or flat bed drier, or mounting down straight away on to mounting board and allowing the photoprint to dry in its own time.

# Aids

### Curves

There are several means of drawing curves besides freehand. The most common are those known as 'french curves'. These are readily available nowadays in cheap plastic forms, either in the flat form or with a bevel on one edge. They come in a multitude of shapes and sizes, but it is an unfortunate fact that they rarely ever perfectly fit the job in hand. One usually finds that this has to be done by means of moving the curve or even using two entirely different curves to get the desired result. A certain amount of hand correction is needed. Another type of curve is flexible which looks rather like a piece of 9 mm ($\frac{3}{8}$ in.) square plastic that can be bent to any desired shape. Unfortunately, this does not take a great deal of sideways pressure and will move if considerable care is not exercised. A very useful type of curve, particularly if it is a fairly lengthy curve that is required, is that of a 60 cm (2 ft) steel ruler. Generally speaking though, unless one has some means of fixing one end of the ruler, it requires a second person to run the pencil or pen around the desired curve. Curves are easily made in fairly thick acetate or thin card. These can be obtained with a certain amount of freedom by simply taking a good sharp blade and making several cuts in as near a curve as one can judge, until the required curve is obtained. When the desired curve has been acquired, it is a simple matter to cut right the way through the card or acetate and then to finish off with a piece of fine glass paper at the same time removing any high spot that may exist. This type of curve is generally only obtainable on long flat curves, and not the short type such as one would obtain with a french curve.

French curves

## Tracing-down medium

To prevent needless use of good paper or board it is good sense first of all to draw out the proposed artwork on flimsy or tracing paper. This is generally done at least twice, particularly with lettering, to get as accurate a tracing as possible. When this final tracing has been acquired, stick it down with two or three short lengths of tape along the longest edge in the desired position on the paper or board. Then take another sheet of flimsy paper approximately the same size as the tracing; on this second sheet lightly rub the surface with a soft pastel (preferably blue in colour if the work is to be done black on white). When it has been sufficiently covered, rub it into the paper well with the fingertips or a piece of cotton wool, removing excess pastel dust. This is then used as one would use a carbon paper, placing it underneath the final tracing with the pastel up against the paper or board. Then taking a hard and finely sharpened pencil, or even better, a steel knitting needle, go over the outline of the drawing, pressing lightly until all the detail has been transferred via the pastel tracing down paper, checking from time to time that important lines have not been left out. When this has been completed remove both the tracing and the tracing down medium which can be used time and time again. The reader may wonder why I have specified blue for black on white work. The reason is that should you fail to clean off the tracing down medium thoroughly when the drawing is completed, it will not 'pick up' under the block or platemaker's lights. Where lettering or illustration is to be drawn on a dark colour background, then yellow, white or very light blue pastel may be used successfully.

A useful addition to the studio and a big time-saver is a *light box*. This can be readily made with the aid of an ordinary bayonet lamp fitting and a 75–100 watt bulb or a small fluorescent tube in a box that is well ventilated, of a size in the region of approximately 15 cm (6 in.) deep and 30 cm × 60 cm (1 ft × 2 ft), depending on the size and nature of the work. Over the top of this box can be placed a suitable thickness sheet of flashed opal glass or ground glass, both of which are generally readily available from glass merchants. Failing this, a sheet of semi-opaque white perspex or a sheet of ordinary glass with a piece of *Kodatrace* or similar material, or even ordinary flimsy or tracing paper taped down on the glass to diffuse the light, will serve reasonably well. The only bother with these materials is that they tend to cockle rather badly with the heat from the lamp. The light box is particularly useful for working directly on to paper instead of making a tracing and then pressing through. An accurate tracing in pencil or, in some cases, direct in pen and ink, may be made, much depends on the subject and its quality of the photograph or object being traced.

### Tints, tones and mechanical tints

Tints such as *Zippatone* and *Letratone* are sometimes used instead of mechanical tints to give depth to drawings laid by the blockmaker which, of course, can be rather expensive and is charged on a square cm (square in.) basis, whereas using *Zippatone* type tints the artist can place down parts of the tint wherever he wishes and with greater accuracy at a much lower cost. Where using these types of tints great care must be exercised in ensuring that they are burnished down perfectly flat, otherwise when photographed by the blockmaker the dots that are not making perfect contact with the paper surface will tend to leave a shadow of themselves, which will be picked up by the camera with undesirable results. Allowances must be made when using these tints for enlargement or reduction; for instance, it is no use whatsoever using a fine dot tint that when on reduction will either begin to break up or to close up and leave a rather muddy effect because the paper on which it is being printed is too coarse. It is as well to check with a book such as the *British Rate and Data* (or *National Guide to Media Selection* in the USA) for all mechanical details regarding size, method and reproduction screen. When using blockmakers' mechanical tints one can either use the old Ben Day tint numbers or specify a tint of a certain screen value. Both *Zippatone* tints and mechanical tints are readily available in positive and negative forms. To differentiate for positive tints for the blockmaker's use, a key line area must first be drawn with as fine a brush or pen stroke as possible, and then the area covered in either blue wash or blue pencil. This clearly defines the tint area. Where a negative tint is required, it is generally the practice to paint the area in either orange or red (both colours will photograph as a solid black), but by referring to the original drawing, the blockmaker may accurately lay a negative tint on the area required.

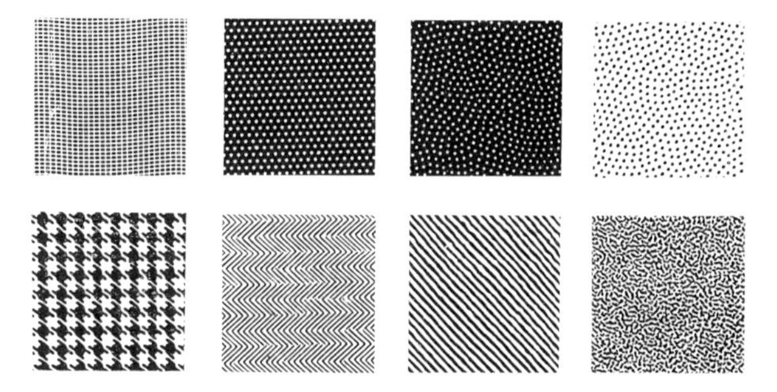

Some examples of Ben Day mechanical tints

*Letratone* examples

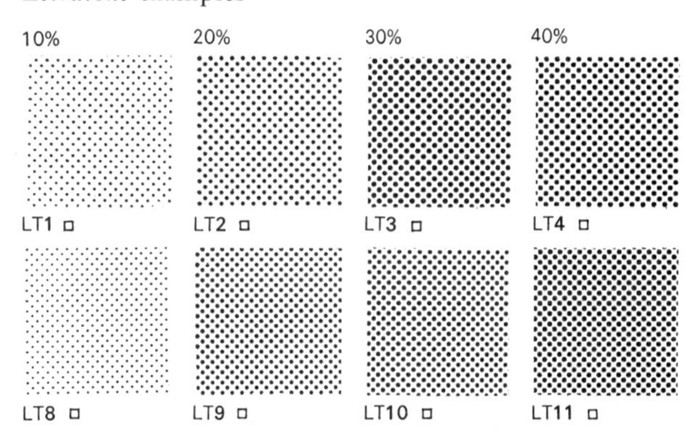


*Letratone* in use

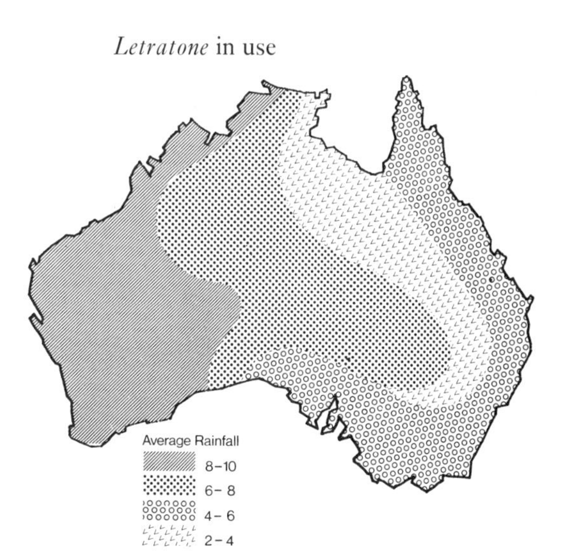

## Graphics for slide projection

These are commonplace requirements in almost general use today.

Many commercial firms, e.g. market research and travel agencies, as well as educational establishments use projections for information and teaching.

The simplest type of projection is the sort that is drawn direct on to a clear film base with felt tipped pens, then laid over a light-box over which is a prismatic lens which can enlarge the image on to a wall or screen.

The next stage is a photodirect method, which is an original being drawn on thin paper (such as layout paper) which is then laid on top of a thin, clear film. Both are then fed through rollers on a small, electrically powered machine (some machines use ultra-violet rays, others use infra-red), and the image is 'burned in' on the film surface, and comes out as semi-opaque on a clear background.

This can be altered chemically into simple colour or colours. Quality of projection, generally speaking, is not very good when compared to normal film projection, but where there are small audiences and rooms, this method is usually found to be adequate.

Transparent, self-adhesive film may be used on areas where colour is required. Most good art shops carry either *Normacolour* or *Letrafilm* gloss, both of which can be cut and peeled away where not required. Care must be taken to exclude grit or dust from contact with the adhesive side of the film, as it is extremely difficult, if not impossible, to remove, and does not only show up on projection but also, if the particles are big enough, cause air bubbles which show.

The use of large slabs of colour on a block graph

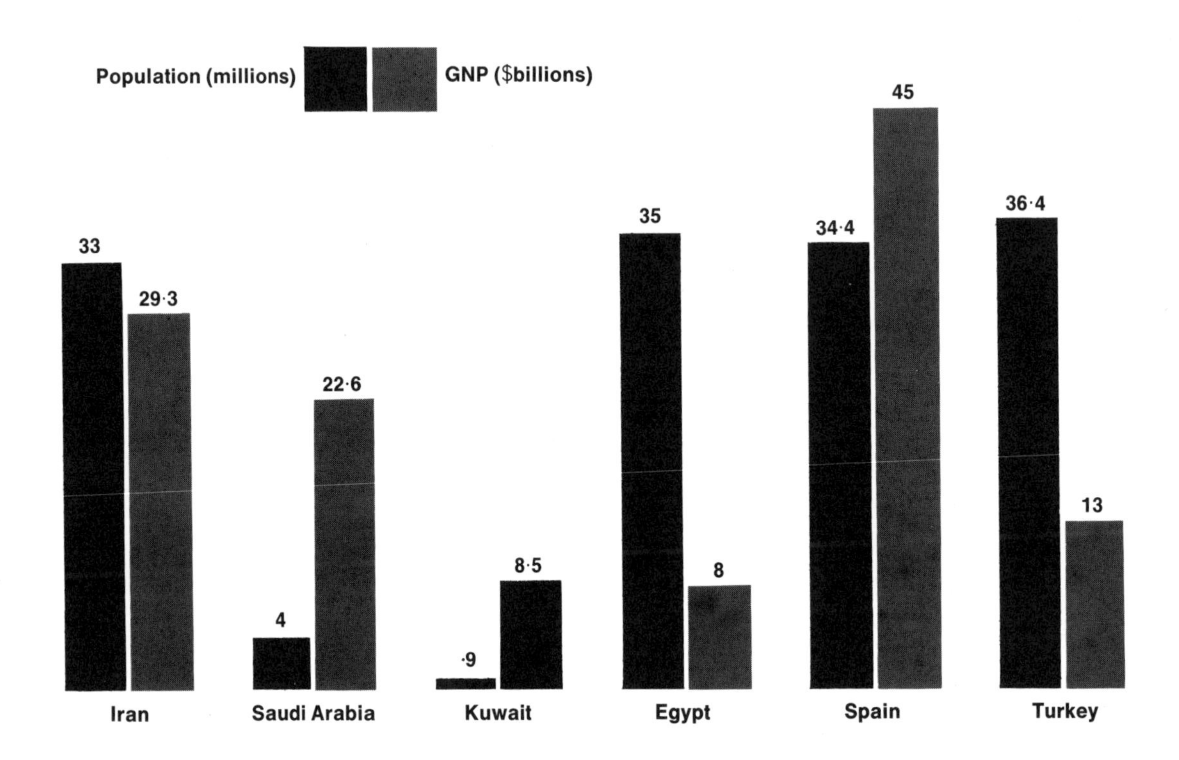

### 35 mm 'Trannies' and 'super-slides' (57 mm: $2\frac{1}{4}$ in. square)

These are extensively used today, particularly in the field of illustrated talks on subjects that require such things as histograms or pie graphs to be shown.

Method of approach by the artist can be varied considerably, depending on the subject.

Probably the best known is the use of coloured felt-tip pens being used for lettering and illustrations on paper or boards up to Imperial size (510 cm × 790 cm: 20 in. × 30 in.).

Where areas of colour need to be laid (such as on block graphs), masking off can be satisfactorily achieved by the use of clear adhesive tape (*viz Sellotape*).

The best results on such artwork are achieved by using a matt surface paper or board, as the colour lays flatter and is much more controllable.

Where large projections are required the image tends to be 'furry' where felt pens have been used, as on close inspection of the original, the grain of the paper becomes apparent, particularly along edges of letters.

Generally speaking, this is a most successful method of working as the colour range is large.

Alternatively, paint or coloured paper may be used on large areas, the latter being stuck down with rubber gum, paste, or twin-sided tape.

A simple, yet effective method of arriving at first class 'trannies' for projection purposes (if money is available) is as follows:

Artwork is made up with all portions in position to fit either an area of 2:3 or 1:1, allowing reasonable margins all round for when the transparency is mounted in the slide carrier – I would suggest margins of not less than 50 mm (2 in.) on artwork of Imperial size (510 cm × 790 cm or 20 in. × 30 in.) – proportionately to this if larger or smaller artwork is prepared.

This is then photographed on line film, usually of half-plate size or thereabouts. The photographer will be required to supply as dense a negative as possible without loss of definition or quality. This will lose many cut lines where patches are used on the artwork. Any cut lines or spots remaining can be opaqued out in the normal way.

Bearing in mind that the whole of the image is now in negative form, i.e. black letters are now clear film, it is possible by means of *Letrafilm* (gloss), *Normacolour* or Clear coloured acetate to colour required areas to any desired effect.

Where large slabs of colour are required (see illustration) they are prepared on the original as solid black.

When the negative has been treated with the colour film, it is then carefully re-copied under controlled conditions on normal 35 mm or 57 mm: $2\frac{1}{4}$ in. square colour film. The end result is, of course, coloured or white on a black background. It is possible, though, to have coloured backgrounds substituted for the black. If transfer lettering or repro type has been used on the original artwork, the end result can be enlarged considerably when projected.

### Printing ink colours

Every studio should have a book of standard printing ink colours. Much time and expense can be saved if it is possible to specify inks from it. If, however, the colour that has been chosen for the job is not standard and the artist is asking the printer to match as perfectly as possible a colour he has concocted, I would suggest that he mixes it two or three shades lighter and then sprays it with a fixative of gum arabic, or even a strip of *Sellotape* burnished down well, to give a simulated effect of glossy ink. This way, it is easier for the printer to match the artist's requirements. Likewise of course, if the artist is asked to match a printer's ink, he will need to do the same sort of exercise.

### Flip-flops

This oddly named article is now a common piece of graphic art used as a sales aid by advertising agents and salesmen, and sometimes by lecturers and teachers.

Generally speaking, the 'flip-flop' is produced on paper of A2 size which, often as not, is punched along one short edge and bound into a book form by means of a ring binder.

They are invariably simple, and bold in treatment, being used as illustrative sales 'punch-homes' to help the salesman or lecturer emphasize important points.

# Hints

## Work surfaces

Working on surfaces such as acetate can prove to be a little difficult unless certain rules are observed, for instance, where highly finished lettering needs to be done, it is necessary first to draw it very carefully in pencil on paper and then place this under the acetate. Having done this it is necessary to cleanse the surface to be worked upon thoroughly and make sure it is free from grease or finger marks. Some artists use pounce powder which is similar to french chalk. This tends to give a surface to the acetate when rubbed over with cotton wool or the finger tips. Generally speaking though, it is best to have at hand either a piece of ordinary soap or ox gall. Ox gall can be obtained in either liquid or solid form. Whichever is used, the procedure is the same: by mixing a little of one or the other with the colour to be used. The disadvantage of soap is that bubbles may ensue and leave unsightly burst bubble marks in the paint as it dries. The disadvantage of the ox gall (apart from the smell) is that the solid form tends to discolour white or light colours slightly. If, for some reason, it is necessary to pick up the acetate and move it from one place to another, make sure that it is well supported and keep in as flat a state as possible, otherwise the slightest bending or kinking will cause the artwork to flake and peel off.

## Removing rubber gum

Removing rubber gum from artwork such as glazed prints, sprayed or painted backgrounds can prove a little difficult. There is a tendency for the rubber gum to leave drag marks on surfaces if the bungie is used as an ordinary rubber. It is best to pick the rubber gum off in a dabbing motion, finally finishing off with a piece of clean cotton wool or rag. If rubber gum, when used on a background, leaves a stain, try putting rubber gum over the whole of the background, allowing it to dry and then picking off carefully in the same dabbing motion. Prior to using rubber gum on coloured backgrounds, it is advisable to give them a good 'fixing' as an insurance against staining.

## Protection of drawings

Protection of drawings is a very necessary part of artwork preparation. Various media can be used such as gum arabic, clear film, cellulose *Transpaseal* or various fixatives that are readily available from art shops.

*Gum arabic* is most commonly used amongst photo retouchers in a diluted form of 10%–20% gum to 80%–90% water mixture. This is then fed into the airbrush and blown over any areas that have been sprayed. If given sufficient thin coats, a reasonably high gloss can be obtained and good protection ensues. Where it is not desirable to spray any protective coating on artwork, clear film may be used as a semi-permanent cover, that is bound round the back of all four edges. This will need removing when it goes for process work.

*Cellulose* in thin quantities may be used in some instances as a protective coat, but here a word of warning: if using any materials that contain cellulose, great care must be taken, particularly by those of the smoking fraternity.

*Fixatives* must be used as specified by the manufacturer. Some fixatives are only suitable for pencil, charcoal or pastel work, but there is some to cover virtually any surface quite satisfactorily.

Whichever method is used other than clear film, allowances should be made for a changing of colour when sprayed. All colours tend to dry at least two shades darker when sprayed.

## Bleeds

Colours that bleed or grin through other colours are fortunately not too common. Colours such as cyprus green, purple, any of the blue-reds are particularly prone to this, especially where white lettering or illustration is to be drawn directly on to the colour. A special white may be obtained for this purpose that comes from Germany and is called *Deck Weiss* by Schminke, it is probably the most successful of all white paints for good coverage. Unfortunately it tends to be slightly yellow in appearance. Where this is unobtainable, it is best in these circumstances before applying any white, to give the background a good fixing with any of the above coatings. If it is necessary to give the colour so much fixative that it becomes shiny, it is likely that the paint will roll back and gather up in globules. In this case, soap or the ox gall treatment will come into its own once more.

## Masking media

Often, when using *Sellotape* as a masking medium or when tabbing down a tracing on to paper or board, the surface will become roughened or even torn on removal. To prevent this, lightly rub the tacky side of the tape on the edge of the work bench, giving it a matt, slightly opaque look. Used like this, it will do the job adequately and will not disturb the surface of the paper, or indeed paint that it may be applied to. If it is not practical to do this, turn the tape underneath itself for about 13 mm ($\frac{1}{2}$ in.), giving somewhere to start peeling from. Peel back at 180° to itself, slowly but positively along the paper or board surface. Do *not* peel upwards or rapidly as damage will result.

## Softening hard colours

Softening hard colours is relatively simple, particularly if the colours are in palettes or pots. If, however, colours are in tubes, this can be a messy operation, as one must first split open the tube and extract the dried colour from within. Having done this, place the dried up colour between two stout pieces of paper, such as brown paper, and hammer or crush into a reasonably fine powder. Then place in a suitable receptacle that will withstand the boiling water to be poured over the powder, stirring until mixed and of a fine consistency. Where colours have dried out in pots or palettes, the same procedure is followed. A tiny drop of glycerine added to the mixture will help to bind the colour together again.

### Two-way tape (twin sided tape)

Two-way tape is an extremely useful item to have in the studio. It is especially useful where paste up work on type is contemplated, particularly where small single items such as letters or numbers need replacing.

### Removing caps from tubes of colour

A successful method of removing paint tube caps that are struck hard is to apply a match flame to the cap for a few seconds, or immerse in hot water. This works well whether the tubes of colour are water colour, gouache, polymer or oil colours.

Brush ruling with wooden 'finger stick'

## Brushes

If one should have the misfortune to break the handle of a brush, it is quite a simple matter to remove this from the ferrule and replace it with another handle. Cut off the broken handle flush with the brush ferrule and then apply a match flame; after a few seconds the remains of the broken handle will pop out as it shrinks. The new handle can then be inserted easily, and fixed in with glue or cement.

## Brush ruler or finger stick

This is an easy method of ruling lines where a ruling pen is not suitable. Flat curves can be obtained with some degree of accuracy and control. It is also a useful method to employ when painting in areas that have been ruled in to take solid colour, enabling the artist to flow on the paint quickly, easily and accurately. The fingers are used to rest on the ruler or finger stick and the second or third finger nail allows the hand to move smoothly along the edge.

Brush ruling with steel ruler

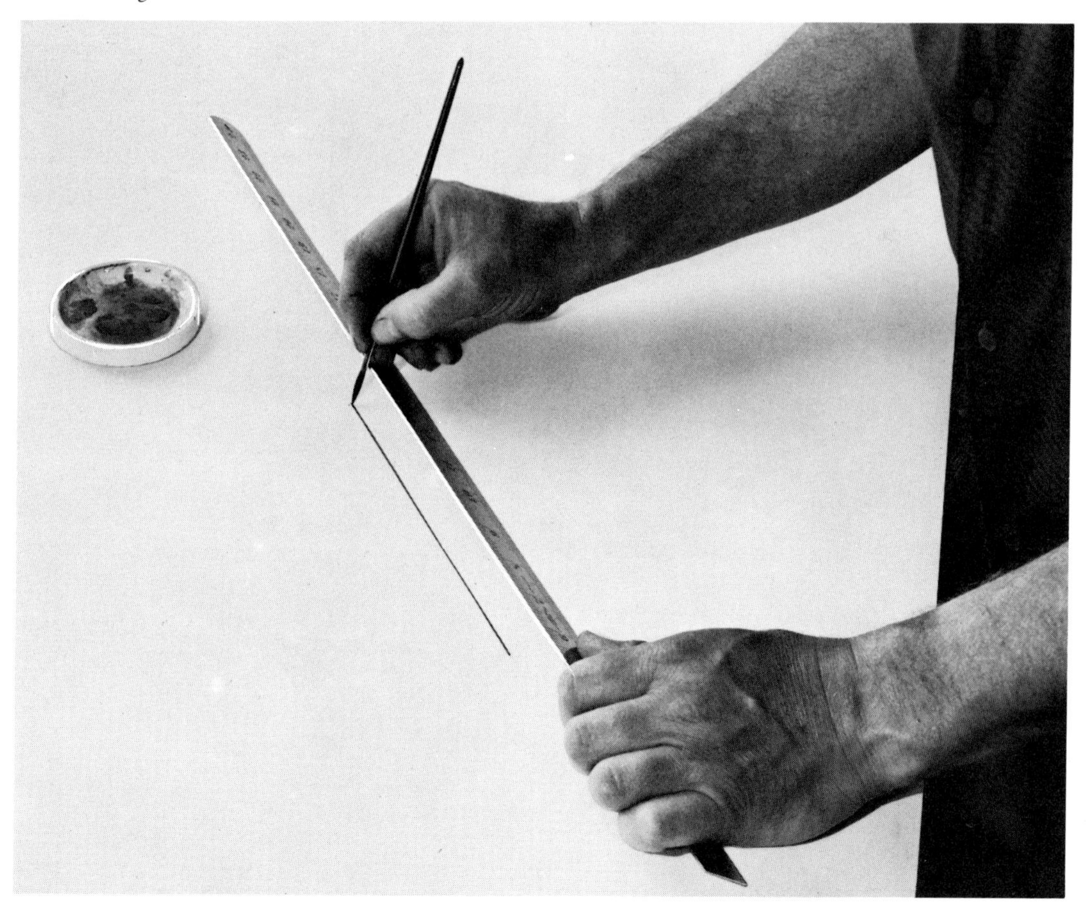

### Mixing colours

A common error where mixing or matching colours is that of trying to lighten colours by means of adding white, instead of working from light to dark. Where a quantity of colour is going to be needed, try to gauge the amount of white paint required to cover the area and then add the colour until the desired shade is obtained. This way, there will be very little waste of paint. Remember, some colours, such as cadmium yellow, can be extremely expensive, therefore it is best to mix a very small quantity as a test run before plunging in with a large quantity of colour that may be wasted.

### Letraset

If a mistake occurs in positioning *Letraset* and the designer is short of time or characters, it is sometimes quite feasible to lay a strip of *Sellotape* over the word or characters that need repositioning and by lightly burnishing the *Sellotape* and then gently peeling it up, the characters will come up almost intact with care. They can then be tabbed down in the required position and left complete with the Sellotape to act as protection. It is a wise precaution, however, when doing this, to trim away the actual edge of the *Sellotape* as very often this has picked up an element of dirt, which is likely to reproduce.

## Stapling pages by hand

Where a long-arm stapler is not available, it is easy to staple books together by hand by setting a pair of hand dividers to the width of the staple and gently pressing them through the folded spine of the booklet and bending them over by hand. This is particularly useful when making up dummy booklets. Before piercing the pages, score and fold them, then pierce from the inside, with the pages held together firmly with a 'bulldog' clip.

## Location of studio

The best position for a studio is facing north, provided that the light is unobstructed. One exception to this rule is where the building opposite (if there is one) has double glazed windows; from practical experience I find that the double glazing acts with a mirrored effect when the sun shines, and can produce direct and dazzling results. Where it is not possible for the studio to face north, it may be found that at some time during the day the use of venetian blinds or cloth blinds may be necessary to prevent dazzle from the paper or board being worked on. If these facilities are unavailable then a sheet of ordinary tissue paper taped on the glass will act very well as a diffuser.

Note 'bulldog' clip holding pages together to ensure that they do not move whilst being pierced

# Typography

Most typefaces can be distorted within the following limits.
Condensing and expanding 12% to 40%
Italicising and backslanting 8° to 22°
Shown here are a few examples of distortions

**Compliments**
CONDENSED 10%

**Compliments**
CONDENSED 40%

**Compliments**
NORMAL

**Compliments**
EXPANDED 10%

**Compliments**
EXPANDED 40%

***Compliments***
ITALICISED

**Compliments**
BACK SLANT

This is a vital section of the designer's art. It is largely a matter of personal taste which typefaces are used but certain types tend to fit certain jobs and seem somehow more sympathetic.

Simple mathematics play a large part in the fitment of type into a given area, and a clear explanation showing how to work out the space required for a given amount of text is given on page 75.

In most large type books Casting Off Charts are given. These are self explanatory and simple to use.

The type used in text matter is generally referred to as *Body Matter* whereas captions and headings are classified as *Display Faces*.

The letters of the alphabet are referred to as *Upper* or *Lower Case*. Upper case letters are those generally known as capitals by the layman, and the lower case as small letters.

Type is generally measured in *Points*. As a rough guide 1 point equals 0.013833 part of an inch. Length of line in type is measured in *Pica Ems* or *Ens*, with 12 points being equal to 1 *Em*, and an *En* is equal to 6 points.

Spaces between lines of type are known as *Leads* (pronounced *Leds*) and is known by compositors as *spacing material*.

A pica em space is equal to 12 points but – only on a type that has a 12 point size body. If a larger or smaller type is employed an *em* space is the same size as the type body. For example, a 16 point face that is indented 1 em at the beginning of a paragraph is not equal to 1 pica em but 16 points, or as the compositor calls it a '16 point *mutton*'. If the type is only on 8 point body then the space would be an 8 point mutton. All very confusing to the layman of course!

We now have the added difficulty of going metric, but whether this will ever become reality where metal types are concerned remains to be seen. As can be imagined the number of type machines and the millions upon millions of metal letters throughout the world will be almost impossible to replace. Not only the type face itself but also the original 'punches' that would have to be remade. The expense would be colossal!

From facts gathered from composing houses, 85% of type for reproduction is *still* measured in *pica ems* and will probably continue to be so for many years. This of course does not apply to the same degree to photo setting or film setting, where it is possible by means of distortion to set type more closely than if it were done in metal. It is not uncommon for film setting whether in text or display faces to be condensed or expanded as much as 15% to 20%. Of course, when such distortions take place the character, colour and weights of each individual letter alter drastically, this is particularly noticeable when letters of a thick and thin construction (examples are shown) are very much condensed, it can be seen from the example that the weights in the letters appear in some instances to be in the wrong places.

With lithography being such a popular method of printing today, film setting is ideally suited for this method of printing. Many books, including this book, are done this way.

## Casting off type

The letters on a standard typewriter occupy the same space horizontally, usually being one of two sizes of typewriter type. Pica type, in which these lines are printed, contains 10 characters to the inch and is called 10-set. Elite, the smaller size, has 12 characters to the inch and is hence called 12-set.*

By measuring a page of typewritten matter with a rule, it is therefore a simple matter to determine the exact number of letters and spaces it contains. By measuring the length of line and counting the number of lines, the number of characters contained on the page can be roughly estimated.

The method is as follows: firstly, rule a vertical line five inches from the left-hand margin, then count the number of full lines in the first paragraph and multiply by 50. Then count the characters outside the vertical line and those in the short lines and add to the total. The result shows total characters and spaces in the first paragraph.

In using the charts, it is necessary to decide upon the typeface to be used, and the size and measure to which the type is to be set, then refer to the appropriate index figure. For example, if this page was to be set in 12 point Baskerville, to a measure of 23 picas, reference to the chart shows that the index figure for 12 point Baskerville, 23 picas, is 52. It is therefore clear at once that this page will make almost the same number of lines in 12 point Baskerville as in typescript. To verify this, divide the number of figures in each paragraph by 52. If the result shows that the copy will occupy too much space for our purpose, try the next size, namely 11 point; the index figure being 57; alternatively, should the copy be required to occupy more space, the index figure for 14 point is taken, namely 48.

* It should be noted, however, that there are on the market typewriters such as the IBM Executive model, whose characters are proportionally spaced, i.e. m is twice the width of n.

## 2 —BELL, MONOTYPE *Series 341*

| Point size | NUMBER OF CHARACTERS TO THE VARIOUS MEASURES FROM 5 TO 36 PICAS. | | | | | | | | | | | | | | | | | | | | | | | | | | | | | | | |
|---|---|---|---|---|---|---|---|---|---|---|---|---|---|---|---|---|---|---|---|---|---|---|---|---|---|---|---|---|---|---|---|---|
| | 5 | 6 | 7 | 8 | 9 | 10 | 11 | 12 | 13 | 14 | 15 | 16 | 17 | 18 | 19 | 20 | 21 | 22 | 23 | 24 | 25 | 26 | 27 | 28 | 29 | 30 | 31 | 32 | 33 | 34 | 35 | 36 |
| 8 | 14 | 18 | 21 | 24 | 28 | 31 | 35 | 38 | 42 | 45 | 49 | 52 | 56 | 59 | 62 | 65 | 69 | 73 | 76 | 79 | 83 | 87 | 90 | 93 | 97 | 101 | 104 | 107 | 111 | 115 | 118 | 121 |
| 10 | 12 | 15 | 18 | 21 | 24 | 27 | 30 | 33 | 36 | 39 | 42 | 45 | 48 | 51 | 54 | 57 | 60 | 63 | 66 | 69 | 72 | 75 | 78 | 81 | 84 | 87 | 90 | 93 | 96 | 99 | 102 | 105 |
| 11 | 10 | 14 | 17 | 19 | 22 | 25 | 28 | 31 | 34 | 37 | 39 | 42 | 44 | 47 | 50 | 53 | 56 | 59 | 62 | 65 | 68 | 71 | 74 | 77 | 79 | 81 | 84 | 87 | 89 | 91 | 94 | 97 |
| 12 | 9 | 12 | 14 | 16 | 19 | 22 | 24 | 27 | 29 | 31 | 34 | 36 | 39 | 42 | 44 | 47 | 49 | 51 | 54 | 57 | 60 | 61 | 63 | 65 | 68 | 71 | 73 | 75 | 78 | 81 | 84 | 87 |
| 14 | 9 | 11 | 13 | 15 | 18 | 20 | 22 | 24 | 26 | 29 | 31 | 33 | 36 | 38 | 40 | 42 | 44 | 46 | 48 | 50 | 52 | 54 | 58 | 60 | 62 | 64 | 66 | 68 | 71 | 74 | 76 | 78 |
| 18 | 7 | 9 | 11 | 12 | 14 | 16 | 18 | 19 | 21 | 23 | 25 | 26 | 28 | 30 | 31 | 34 | 36 | 38 | 39 | 40 | 42 | 44 | 46 | 48 | 50 | 52 | 53 | 54 | 56 | 58 | 60 | 62 |

## 3 —BEMBO, MONOTYPE *Series 270*

| Point size | NUMBER OF CHARACTERS TO THE VARIOUS MEASURES FROM 5 TO 36 PICAS. | | | | | | | | | | | | | | | | | | | | | | | | | | | | | | | |
|---|---|---|---|---|---|---|---|---|---|---|---|---|---|---|---|---|---|---|---|---|---|---|---|---|---|---|---|---|---|---|---|---|
| | 5 | 6 | 7 | 8 | 9 | 10 | 11 | 12 | 13 | 14 | 15 | 16 | 17 | 18 | 19 | 20 | 21 | 22 | 23 | 24 | 25 | 26 | 27 | 28 | 29 | 30 | 31 | 32 | 33 | 34 | 35 | 36 |
| 10 | 13 | 16 | 19 | 22 | 26 | 29 | 32 | 35 | 38 | 41 | 44 | 48 | 51 | 54 | 57 | 61 | 64 | 67 | 70 | 73 | 76 | 79 | 82 | 85 | 88 | 91 | 95 | 99 | 102 | 105 | 108 | 111 |
| 11 | 11 | 14 | 17 | 20 | 23 | 26 | 29 | 32 | 35 | 38 | 40 | 43 | 46 | 49 | 52 | 55 | 58 | 61 | 64 | 67 | 70 | 73 | 76 | 79 | 81 | 83 | 86 | 89 | 92 | 95 | 98 | 101 |
| 12 | 10 | 13 | 16 | 18 | 21 | 24 | 27 | 30 | 32 | 35 | 38 | 40 | 43 | 46 | 48 | 51 | 54 | 57 | 60 | 63 | 65 | 67 | 70 | 73 | 76 | 79 | 81 | 83 | 86 | 89 | 92 | 95 |
| 13 | 9 | 12 | 14 | 16 | 19 | 22 | 24 | 27 | 29 | 31 | 34 | 36 | 39 | 42 | 44 | 47 | 49 | 51 | 54 | 57 | 60 | 61 | 63 | 65 | 68 | 71 | 73 | 75 | 78 | 81 | 84 | 87 |
| 14 | 9 | 11 | 13 | 16 | 18 | 21 | 23 | 26 | 28 | 30 | 33 | 35 | 37 | 40 | 43 | 45 | 47 | 49 | 52 | 55 | 57 | 59 | 60 | 63 | 66 | 69 | 71 | 73 | 75 | 77 | 80 | 83 |
| 16 | 8 | 10 | 12 | 13 | 16 | 18 | 20 | 21 | 24 | 26 | 27 | 29 | 31 | 33 | 35 | 38 | 40 | 42 | 43 | 44 | 47 | 50 | 52 | 54 | 55 | 57 | 58 | 60 | 62 | 64 | 66 | 68 |

**Simulation of body matter or text**

'Greeking'
(brush, pen or pencil)

Z10 Soft anb mjd ligmtrane
aexbog wrj sruym
B, C & D anq 40 crs – 39 nqw

'Tramline' copy
(pen or pencil)

'Scribble' copy
(pen or pencil)

Square cut felt pen copy
(sometimes done in black or grey,
depending on boldness required)

Self-adhesive 'Body Type' made by Letraset Limited (black on white or white on black)

Lorem ipsum dolor sit amet,
voluptat. Ut enim ad minim ve
vel eum irure reprehenderit in
iusto odio dignissim ducim qui

**BT5** 10 point leaded 3 points

Lorem ipsum dolor sit ame
aliquam erat voluptat. Ut eni
consequat. Duis autem vel
pariatur. At vero eos et acc

**BT11** 10/11 point leaded 3 points

Lorem ipsum dolor sit ame
aliquam erat voluptat. Ut e
commodo consequat. Duis
eu fugiat nulla pariatur. At

**BT3** 12 point set solid

Lorem ipsum dolor sit
labore et dolore magna
suscipit laborios nisi
voluptate velit esse nihi

**BT9** 12/13 point set solid

Examples of good and bad spacing of transfer letters

BLOOD Bad spacing WATER

BLOOD Good but wide WATER

BLOOD Better with nipped off serifs on T and E and shortened bottom bar of 'L' WATER

The A T and E touching WATER

Too tight
berry

Too wide
berry

Better
berry

Fancy hand drawn lettering originally in full colour

Kenya Pure

Outline lettering

Coffee

Hand drawn lettering adapted from
Albertus type

Th R H F T

y l i h m k

x j n p f

s q u a e

C S fl

Part of hand drawn alphabet for TAP airlines. Note parallel guide lines for ease of aligning letters when using photoprints to make up captions

W I A ! D

ff b r z

'X' HEIGHT.

g t c o P

d w v ? g M

B L 72 A

Some specimens of typefaces from a typesetters synopsis – showing range of point sizes

**Rockwell Heavy Condensed**
14 18 24 30 36 42 48 60 72

**Rockwell Extra Heavy**
14 18 24 30 36

ROCKWELL SHADOW
18 24 30 36 48

Roman Compressed
18 24 30 36 48

**Roman Sphinx**
48

Saxon Black
12

Scotch Roman
8 9 10 11 12 14 18 24

*Scotch Roman Italic*
8 9 10 12 18

SLIMBLACK
30 60

Spectrum
16D/18 24D/30

*Spectrum Italic*
24D/30

Standard Light Extended
6 SF 6 LF 8 10 12 14 18 24 SF 24 LF 30

Standard Extended
6 SF 6 LF 8 10 12 14 18 24 SF 24 LF 30

**Standard Medium**
10 18 24 SF 24 LF 30 42

Studio
24

*Temple Script*
24

THORNE SHADED
24

Times
5 6 7 8 9 10 11 12 14 18 24 30 36 42 48 60 72

*Times Italic*
5 6 7 8 9 10 11 12 14 18 24 30 36 42

**Times Semi-Bold**
8 9 10 12 14 18 36

**Times Bold**
6 7 8 9 10 11 12 14 18 24 30 36 42 48 60 72

***Times Bold Italic***
6 8 10 12 14 18 24 30 36

TIMES TITLING
14 18 24 30 36 42 48

**TIMES BOLD TITLING**
18 24

Marwick & Paulig Ltd
Superline
Drawing Board

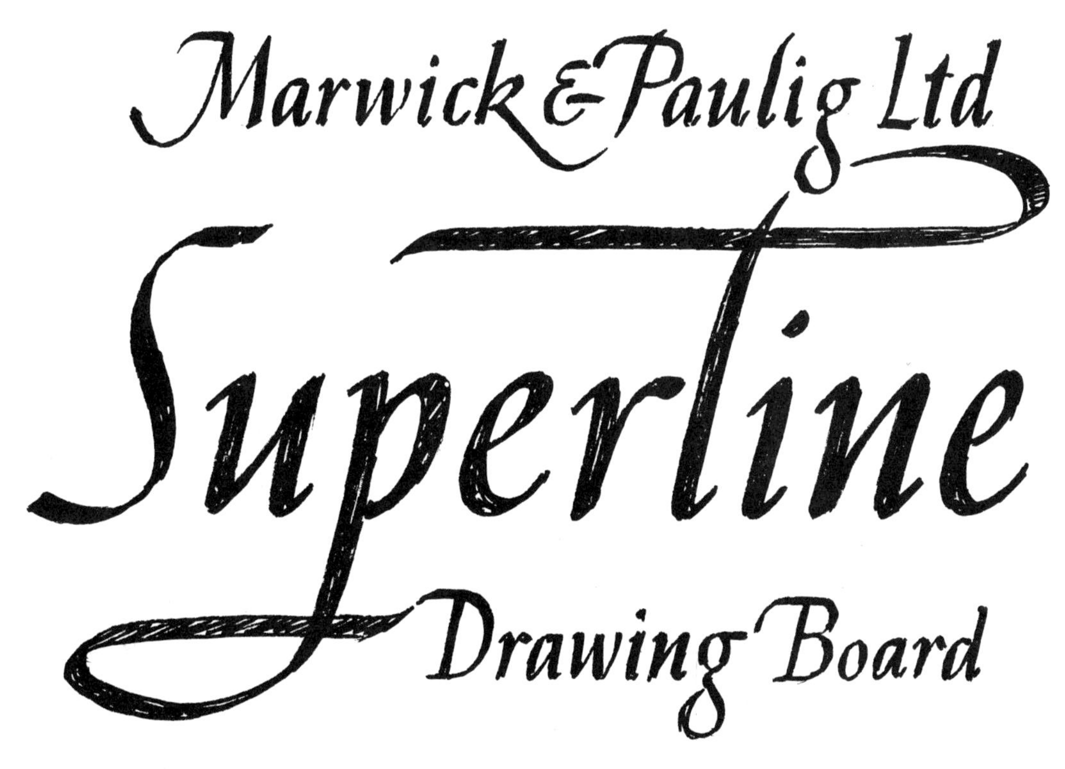

Rough one colour design

Printed version down to size

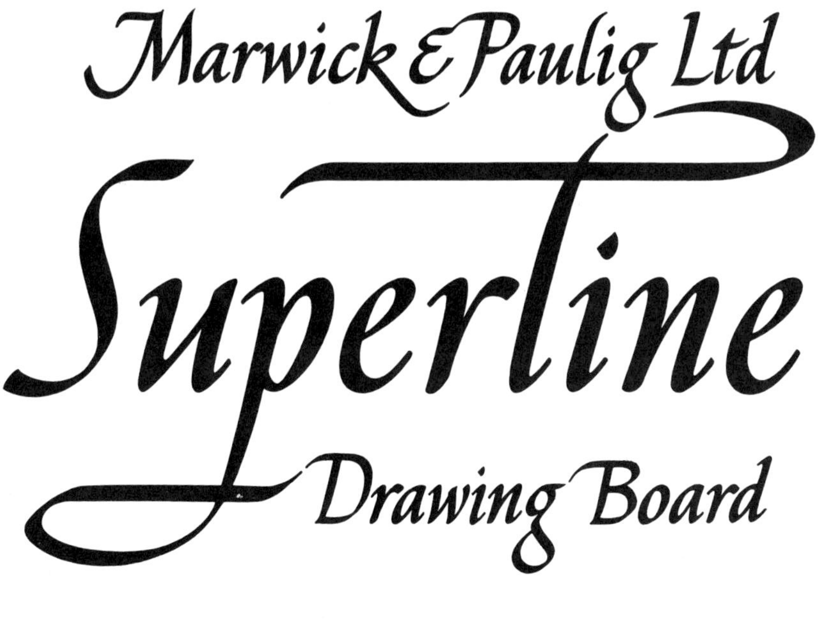
Marwick & Paulig Ltd
Superline
Drawing Board

Finished artwork

Marwick

Supe

Dra

Paulig Ltd

rline

ving Board

Original reproduction size of advertisement $14\frac{7}{8}$ in. × $10\frac{3}{4}$ in.

NB Sizes of advertisements are always quoted depth first and width last. In the USA this is *vice versa*

Scraperboard illustrations

Welcoming new
Player's Nº6 Classic
–a lot more to offer 23½p* for twenty
6 vouchers
* Recommended price

PSC 2F

PLAYER'S
Nº6
CLASSIC
Filter Virginia

FREE!
Get a FREE packet of twenty Player's Nº6 Classic plus FREE Bonus vouchers – just send us the vouchers from 7 packets of 20 Player's Nº6 Classic. Details in pack or ask your retailer.
Offer closes 31st December 1973, and is restricted to smokers aged 18 years and over resident in the UK.

EVERY PACKET CARRIES A GOVERNMENT HEALTH WARNING

Original reproduction printed on A4 paper

# MENU

# THE MOBY DICK

WHALEBONE LANE

CHADWELL HEATH, ESSEX. TEL. SEVEN KINGS 9524

# Glossary

| | |
|---|---|
| Acetate | Clear film used for overlay work, sometimes called *clear cell* |
| Airbrush or Airgun | A small pen-shaped spray gun working on compressed air. Used for obtaining smooth tones on retouching or spraying backgrounds |
| Align | To arrange in a line |
| Ampersand | & an abbreviation for 'and' |
| Art paper | Smooth, hard coated paper used for printing; coated with china clay and polished |
| Artwork | Presentation drawings, lettering, paste up or photographs for reproduction |
| Ascender | The stroke which ascends above the main body of the character as in 'h' |
| Autone prints | Colour or metallic-based photoprints |
| Barbola paste | Embossing medium |
| Batter | A damage to the printing surface of printing blocks or type characters |
| Beard | The space from the bottom of the x-height of the type to the front of the shank |
| Ben Day tints | A method of tint laying on negative, positive or metal print. Invented by Benjamin Day |
| Bleach out | An underdeveloped bromide print used as a basis for a line-drawing. After drawing the bromide print is bleached away |
| Bleeding off | The allowance on a drawing or printing plate which extends the plate beyond the trimmed and finished size to ensure a clean cut-off. Usually 3 mm ($\frac{1}{8}$ in.) |
| Blind embossing | A design stamped on to paper or board without the use of gold leaf or ink |
| Blocks | Letterpress. A halftone, line or duplicate printing plate |
| Blow up | An enlargement, usually photographically |
| Body | In typesetting, term used to describe the size of the body of type, e.g. 10 pt |
| Body colour | Opaque colour |
| Boldface | A heavy type used for emphasis on headings, etc |
| Brief | Instructions from client to designer |
| Bristol board | A top quality rag paste-board with a high smooth finish |
| Bromide prints | Photographic prints |
| Bromfoil prints | Metal sandwiched photographic prints |
| Brush ruling | Method of ruling |
| Caliper | Thickness of paper or board expressed in thousandths of an inch |
| Camera Lucida 'Lucy' | An original method of projection |
| Caps and small caps | Two sizes of capital letters on the same body size in the same fount |
| Caption | Description of illustration |
| Cartridge | A strong opaque drawing or printing paper |

| | |
|---|---|
| Cartouche | A fancy border |
| Casting off | The calculation of typewritten copy to be set in type |
| Character | Letter or numeral |
| Cold setting | Type in the form of film or separate metal letters |
| Colour separation | Line or tone |
| Colour swatch | Colour match indication |
| Compose | Type arrangement |
| Condensed type | Narrow measure letters |
| Contact print | Same size as negative |
| Copy | Any material provided by the client (manuscript, photographs or drawings) to be used in the production of printed matter |
| Copy negative | Negative made by copying original flat print or artwork |
| Cover paper | General term for heavy, coloured papers used for covers of artwork catalogues, brochures, etc |
| Creasing | Impression made by a rule to break the grain of the paper to facilitate folding |
| Cross line screen | Standard halftone used for halftone work. Made by cementing together two ruled glass screens at 45° |
| Cross hatch | Criss-cross patterns and textures usually made with pen and ink |
| Crown | Standard size of printing paper 38 cm × 50 cm (15 in. × 20 in.) |
| Cut out | In process-engraving, a halftone printing plate with the background cut away to the outline of the subject |
| Cutters | Hand-made shapes for cutting out intricate printed labels |
| Cylinder press | A press on which the type form is flat but the printing action is made against a revolving cylinder |
| Deckle | The rough, natural edge on paper |
| Deep etch | In lithography, a plate for long runs where the image is slightly recessed below the surface of the plate |
| Deep etching | The removal of halftone dots from specified areas of the printing plate |
| Descender | That part of the type which extends below the main body of the character, e.g. 'p' |
| Didot | Type measurement used mainly on the Continent of Europe |
| Die stamping | An intaglio printing process from a steel die with a stiff ink, which leaves the printing surface in relief. Used mainly for letterheads |
| DIN sizes | Deutsche Industrie Normie paper sizes. The basic international paper size |
| Dot for dot | A same size line reproduction of a good quality halftone art pull or proof where the original has been lost or damaged. The coarser the halftone the better will be the quality of copy |

| | |
|---|---|
| Double printing | In process engraving, two exposures from separate negatives on one piece of metal prior to etching |
| Dow etch | Method of powderless etch patented by the Dow Chemical Corporation for etching without side loss or underbite |
| Dragons blood | Name given by the trade to a red bituminous powder used for dusting the side walls of a line plate during etching to prevent the acid underbiting the printing surface |
| Dummy | A paste-up or pattern of a proposed piece of printing |
| Dummy boxes | Mock-ups of the real thing |
| Duotone | A two-colour halftone set produced from a single colour original |
| Duplicate blocks | Sets of blocks all exactly alike |
| Dye transfer | Colour print process (photographic) |
| Electro | A duplicate letterpress printing plate made by electrolytically depositing copper on a mould taken from an original plate or type and backed with a lead alloy |
| Em | The width of the body on which the lower case 'm' is cast in a fount of type. If not prefixed by type size (e.g. 10 pt em) it is taken to be a pica em (12 pt) |
| Embossing | Impressing characters or a pattern on paper or board to give a raised surface |
| En | Half the width of an em. The basic unit for casting off copy for typesetting |
| Enamel paper | A paper coated on one side and polished to give a high gloss finish |
| Engraving | Cutting into face of metal block by hand |
| Face | That part of the type or plate which makes contact with the paper |
| Felt pens | Felt tipped pens with spirit type ink or dye in body |
| Filter | In process engraving, the coloured gelatine or glass placed in front of the camera lens to separate colour |
| Fineline | Highly detailed block |
| Fixative | Thin cellulose or shellac to act as a protection to drawings |
| Flat bed press | A printing machine which has the printing surfaces in a flat plane |
| Flexi chrome | A gelatinous colour print produced first as black and white and stained to colour. The gelatine absorbs stain in ratio to the density of the original |
| Flexographic | Printing by letter press from flexible (usually rubber) plates |
| Flong | A specially prepared paper used for taking moulds for stereotyping |
| Flour paper | Fine glass paper |
| Flush (left or right) | A term indicating that typematter to be set to line up at the left or right |
| Fount | A complete family of letters, characters and numerals of one size and style of type |

| | |
|---|---|
| Four colour | Artwork or blocks reproducing a full colour effect |
| Frisket | Thin, semi-transparent, glazed paper |
| Furniture | Typesetting metal more than 2 pica ems in width, used to fill blank spaces in the frame |
| *Gillac* | Aerosol fixative spray |
| Gravure | Method of printing, see Photogravure |
| Greeking | Pseudo lettering on roughs |
| Glazed prints | Photographic prints with hard, shiny surface |
| Goobungie | *Cow* gum rubber |
| Gouache | Opaque watercolour |
| Gum arabic | Protective spray gum |
| Gum strip | Brown paper with one sticky side |
| Gutter | The margin between the sides of two pairs of pages |
| GSM (paper weight) | Grammes per square metre |
| Hairline spacing | Very fine letterspacing |
| Halation | The reflection of scattered light back through the photographic emulsion producing a diffused halo around the bright areas |
| Hot metal setting | Moulded type |
| Halftones | A print giving the optical illusion of continuous tone by small dots of varying size. The ink film on all dots is constant but the eye mixes the white from the paper with the varying size dots and produces varying greys and an illusion of continuous tone |
| 'H' height | Letters measured over the 'ascender' |
| Highlight | The lightest tonal value of a halftone |
| Hot press paper | Smooth, toothless paper |
| Intaglio | Any printing process from a recessed image, e.g. gravure, copper-plate, die stamping, etc |
| Intertype | A mechanical typesetter which casts the characters in one line as a complete slug. See Linotype |
| Italic | Slanted letters |
| Justify | The arranging of type and word spaces to give lines of equal length |
| Kern | Any portion of type character which overhangs the body |
| Key drawing | An outline drawing from which can be prepared colour printing plates by set-offs or keys without recourse to colour filters in the camera |
| Key-lines | Lines on a drawing to indicate to the process worker areas for tint-laying, painting up, etc |
| *Kodatrace* | Synthetic transparent material used for overlays of drawings |
| Landscape | A sheet or book with the long edges at head and foot |
| Layout | Rough design in art – position of subjects in print |
| Leaders | Dashes or dots in composition used to guide the eye across the page to a word or figure |
| Letterpress | The process of printing from a raised image |

| | |
|---|---|
| Letter space | The fine spacing between characters in a word to give an optical balance or to improve legibility |
| *Letraset* | A system of self-adhesive type characters which can be attached to designs or artwork |
| Ligature | Two or more letters joined together on one body, i.e. fi, fl, ff, etc |
| Line artwork | Artwork of a purely black and white nature with no mid-tones |
| Line block | A relief printing plate produced from a line drawing without the use of a screen |
| Line/tone combine | A letterpress printing plate containing halftone and line printing arranged in such a manner that the plate is etched first as a halftone then as a line-block |
| Linotype | A mechanical typesetter which casts characters in one line as a slug |
| Lithography | Printing from a flat surface of stone or metal. The image areas on the plate being protected by a water repellent ink and the non-image areas protected by a film of water |
| Logotype | A unique arrangement of letters used as a trade design not necessarily registered |
| Ludlow | A slug-casting system from handset matrices. Most common use is for newspaper headlines |
| Lower case | The small letters (a b c d, etc) as distinct from capital letters (A B C D, etc) |
| Make up | Process of assembling elements for reproduction |
| Make ready | In letterpress printing, the preparatory work involved on the machine and printing forme to obtain correct printing impression over the whole area |
| Mask | In advertising photography, an opaque cut-out overlay which masks out the unwanted portion of a photograph during process camera operations. In process engraving, a photographic negative or positive used for colour correction |
| Masking medium | *Sellotape*, *Frisket*, etc |
| Matrix | A mould for casting type or a mould for papier mâché, plastic, rubber, etc, for duplicate plates |
| Measure | Width or depth of type area in 12 pt (pica) ems |
| Moiré | Pattern formed by conflict of angles of process screen and lines or patterns in the original |
| Monochrome | An original in one colour only |
| Monotype | A mechanical system for casting single type characters |
| Montage | Fitting and faking photoprint images |
| Mountant | Paste, gum, etc |
| Negative | A photographic image in reverse tonal value to the original copy |
| Negative tint | White dots or pattern on solid ground |
| Newsprint | A cheap paper, made mainly from mechanical wood, with characteristics suitable for newsprinting |

| | |
|---|---|
| Offset lithography | The lithographic printing method where the image is printed to a rubber plate and transferred (offset) to paper |
| Opaque | Photographic paint used on negatives for blocking out unwanted portions |
| Original | A term applied to a photograph or drawing used for reproduction. Also used to describe a printing plate used as a master plate for duplication |
| Orthochromatic | Does not register blue on the film |
| Overlay (Koda) | Overlay for stripping. A flap over the original drawing indicating areas for tints, etc or giving special localised instructions to the process worker |
| Overprint | Photographic – two images exposed on same paper one over the other<br>printing – two images printed one over the other |
| Panchromatic | Sensitive to all colours (film) |
| Paste up | Preparation of artwork by designer putting all elements in position |
| Photogravure | An intaglio printing process where the printing image is produced photographically |
| Photolithography | Lithographic process where the printing image is produced photographically |
| Photosetting | Type that is reproduced photographically and may be distorted and spaced at will |
| Photostat | Paper negative instead of film negative with positive on paper |
| Pica | A standard unit of measurement in the point system 1 pica = 12 pt |
| Point | The smallest unit of measurement in the point system |
| Polyester film | Synthetic film such as *Kodatrace* |
| Portrait | An upright page with longest edges at sides |
| Positive | A photographic image on glass or paper which corresponds to the original copy (reverse of negative) |
| Positive tint | Series of dots or patterns on white ground |
| Poster colours | Coarse, cheap type of water-based body paint |
| Pounce powder | French chalk based material used for absorbing greasy marks on paper, etc |
| Primary colours | In printing inks, yellow, magenta and cyan; in artists colours, red, yellow, blue |
| Process plates | A set of printing plates made in halftone to produce a wide range of colours and shades. Usually three or four colour process, yellow, magenta, cyan and black |
| Presentation roughs | Roughs for client presentation – usually of a high standard |
| Process colours | Yellow, magenta, cyan and black |
| Process engraving | Term used to describe engraving by photo-chemical or photo-mechanical means to produce letterpress printing plate |

| | |
|---|---|
| Progressives | A set of proofs showing each plate of colour set printed in correct colour and sequence |
| Proofs | Pulls taken from the type or blocks. Usually prefixed to indicate the type of proof, e.g. galley proofs, press or machine proofs, progressive proofs, etc |
| Pull | Art paper reproduction |
| Quartertone | Line block reproduced from 'doctored' halftone |
| Quire | One twentieth part of a ream |
| Ream | Normally in printing papers, 500 sheets. In some writing papers, 480 sheets |
| Register marks | Used to correct relative position of two or more printings on same sheet |
| Repro type | High quality reproduction type for use in studios on artwork |
| Re-touching | Photographic correction and improvement |
| Reversal | White image from black original or vice versa |
| Reproduction proof | A proof taken from type for the purpose of photographic reproduction |
| Roman | Upright letters |
| Rotary | Printing from curved plates |
| Routing | Cutting away the non-printing areas of a letterpress plate to bring them below printing height |
| Photogravure (gravure) | Method of printing |
| Sans serif | Letters without serifs, usually referred to as 'sans' |
| Scamp | Rough of a very basic nature |
| Score | To impress a mark with a rule to simplify paper or board folding |
| Scraper board | Heavily coated board also called 'scratch board' |
| Screen process printing | A method of printing using a squeegee to force ink through a stencil supported by a fine mesh fabric or metal screen |
| Separation negs | A set of negatives produced through colour filters from a coloured original |
| Set off | A condition where ink is transferred from one sheet to the back of another accidentally |
| Serif | The short cross lines at the ends of the main strokes of certain type faces |
| Slug | A complete line of type cast in one piece |
| Splatter brush | Toothbrush loaded with colour, then splattered by rubbing the hairs one way with a knife |
| Squared up tones | Blocks or plates of squared shape |
| S/S | Same size as the original |
| Stanley knife | Heavy craft knife |
| Stereo | A duplicate printing plate made by casting in type metal |
| Stipple | Dotted pen texture |
| Strawboard | A board made mainly from straw fibres |
| Stripping | The removal of photographic emulsion from its support to assemble with others on another support |

| | |
|---|---|
| Stripping guide | Guide for showing where certain positions of artwork are to be laid on plates or blocks |
| Swatch | A colour specimen |
| T-squares | Wooden, metal or plastic T-shaped straight edges used in conjunction with a true-edged board |
| Tints | A solid area of printing plate to be printed in a lighter shade of ink, or a regular tone produced by a dot formation to give the appearance of a lighter tone |
| Tracing medium | Pastel coated paper |
| Transparency (trannies) | A photographic positive for viewing by transmitted light |
| Transpose | To exchange the position of one letter, word, group of words or illustration with another letter, word, group of words or illustration |
| Thermographic | The process of dusting freshly printed sheets with resinous powder. When heated the fused powder forms a raised surface similar to die-stamping. Mainly used for letterheads and brochure covers |
| Typography | The art of using type |
| Two-way tape (twinstick) | Double-sided sticky tape |
| Unmounted blocks | Blocks not mounted on wood |
| Vignette | A halftone with the background gradually fading away and blending into the surface of the paper |
| Vellum | Prepared calfskin. Also used to describe the finish on some writing papers |
| Wash drawing | A monochrome drawing where tones are formed by washes of grey and black |
| Water mark | A design impressed into the paper web as it passes under the dandy roll in a wet state |
| Water colour | Translucent water paint |
| Web offset | Method of printing using a continuous roll of paper rather than sheets |
| Wet plate | A photographic plate prepared in the process shop and used in the camera whilst wet |
| Wood cut or wood engraving | Hand carving of illustration in box wood |
| Work and tumble | Printing the second side of the sheet by turning over from gripper to back |
| Wrong fount | A type character which does not correspond to the type face being set |
| Work and turn | Printing the second side of the sheet by turning it from left to right, using the same gripper edge |
| X height | The height of the lower case without ascenders or descenders. Different type faces have different 'x' heights |
| *Xylene* | A solvent which removes unwanted felt pen marks |
| Zinco | A letter press printing plate made on zinc |
| *Zippatones* | A range of self-adhesive tint media which can be cut and attached to drawings to give special tones and tints on the finished plate |

# Suppliers list

The majority of items mentioned in this book may be obtained from any good local artist's colourman or stationers. Specific products and catalogues available on request from

Letraset UK Limited
17–19 Valentine Place
London SE1 8QW

Letraset USA Incorporated
33 New Bridge Road
Bergenfield
New Jersey 07621

*Letraset, Letracote, Letrafilm, Instantex*

Winsor and Newton Limited
57 Rathbone Place
London W1

Winsor and Newton Incorporated
555 Winsor Drive
Secaucus
New Jersey 07094

*Aerosol sprays, solvents, papers and all art materials*

A West and Partners Limited
684 Mitcham Road
Croydon
Surrey

*Manufacturers of UNO Drafting Film and pounce powder*

Focus Photoset Limited
134 Clerkenwell Road
London EC1

*Photosetting*

Impact Typesetters
65–69 Leonard Street
London EC2

*Typesetting for reproduction*

Marwick and Paulig Limited
Legion Works
Kimberley Road
Kilburn, London NW6

*Art board manufacturers and paper suppliers*

London Graphic Centre
107–115 Long Acre
London WC2

*Artists' supplies*

Copies of *British Rate and Data* (BRAD) or *The National Guide to Media Selection* should be available in reference libraries if not already on your shelves. Information regarding these publications from

British Rate and Data Publications
30 Old Burlington Street
London W1X 2AE

The Standard Rate and Data Service Publishing Inc
5201 Old Orchard Road
Skokie
Illinois 60076